M000250106

RUSSIA'S
POLICY CHALLENGES

RUSSIA'S
POLICY CHALLENGES

SECURITY, STABILITY, AND DEVELOPMENT

STEPHEN K. WEGREN
EDITOR

M.E. Sharpe
Armonk, New York
London, England

Copyright © 2003 by M. E. Sharpe, Inc.

All rights reserved. No part of this book may be reproduced in any form
without written permission from the publisher, M. E. Sharpe, Inc.,
80 Business Park Drive, Armonk, New York 10504.

Library of Congress Cataloging-in-Publication Data

Russia's policy challenges: security, stability, and development /
[edited] by Stephen K. Wegren.
    p. cm.
Includes bibliographical references and index.
ISBN 0-7656-1079-5 (alk. paper)—ISBN 0-7656-1080-9 (pbk.: alk. paper)
    1. Russia (Federation)—Politics and government—1991–2. National security—
Russia (Federation) 3. Russia (Federation)—Social policy. 4. Russia (Federation)—
Economic policy—1991– I. Title: Title on CIP data: Russia's policy changes. II. Wegren,
Stephen K., 1956–

JN6695 .R868 2003
320′.6′0947—dc21

2002030925

Printed in the United States of America

The paper used in this publication meets the minimum requirements of
American National Standard for Information Sciences
Permanence of Paper for Printed Library Materials,
ANSI Z 39.48-1984.

BM (c) 10 9 8 7 6 5 4 3 2 1
BM (p) 10 9 8 7 6 5 4 3 2 1

This volume is dedicated to the next generation
of students, who hopefully will study Russia, understand its
problems and potential, and help it become a better place to live.

Table of Contents

List of Tables, Figures, and Map

TABLES

FIGURES

MAP

About the Editor and Contributors

Stephen K. Wegren is associate professor of political science at Southern Methodist University. He is the author or editor of six books and dozens of articles on the politics and economics of Russia's transition, including most recently *Rural Reform in Post-Soviet Russia* (2002); "Winners and Losers in Russian Agrarian Reform," *Journal of Peasant Studies* (2003); and *Searching for Agrarian Capitalism in Russia* (forthcoming).

Mikhail A. Alexseev is associate professor of political science at San Diego State University. He is the author of *Without Warning: Threat Assessment, Intelligence, and Global Struggle* (1997) and *A Federation Imperiled: Center-Periphery Conflict in Post-Soviet Russia* (1999). His current research focuses on ethnic hostility and public opinion concerning Chinese migration into the Russian Far East.

Vladimir R. Belen'kiy is professor and director of the Institute for Land Relations and Land Planning in Moscow. He has authored or coauthored over 200 books and articles on social development of villages, land relations, and the land market in Russia.

Herbert J. Ellison is professor of history and international studies at the University of Washington. He has written extensively on Russian foreign policy, particularly relations with East Asia and Western Europe. He is currently completing a book manuscript on the Yeltsin and Putin eras that provides extensive treatment of the transformation of Russia's international role following the collapse of the Soviet Union.

Gregory Gleason is professor of political science at University of New Mexico. His research interests focus on political relations with Near Abroad and the political economy of development in Central Asia.

Timothy Heleniak is adjunct professor at Georgetown University and a demographer at the World Bank. He has published widely on migration and demographic problems in Russia. The research for his chapter was completed while he was a research fellow at the Kennan Institute, Woodrow Wilson International Center, in Washington, DC.

Dale R. Herspring is professor of political science at Kansas State University. He is the author or editor of nine books and dozens of articles on the Soviet Union, Germany, and Poland. His most recent book is titled *Putin's Russia: Past Imperfect, Future Uncertain* (2002).

Debra Javeline is assistant professor of political science at Rice University. Her research interests focus on political behavior. Her publications include *Protest and the Politics of Blame: The Russian Response to Unpaid Wages* (2003) and "The Role of Blame in Collective Action: Evidence from Russia," *American Political Science Review* (2003).

Christopher Marsh is associate professor of political science at Baylor University. His research interests focus on communist studies and transition politics in Russia and China, about which he has published numerous articles and books. His most recent books include *Russia at the Polls: Voters, Elections, and Democratization* (2002) and *Civil Society and the Search for Justice in Russia* (2002).

James R. Millar is professor of economics and international affairs at George Washington University. He is former director of the Institute for European, Russian, and Eurasian Studies at GWU. He was also the editor for *Problems of Post-Communism.*

Joel C. Moses is professor of political science at Iowa State University. His research focuses on regional politics in Russia. His recent publications include *Dilemmas of Transition in Post-Soviet Countries* (2002) and "Political-Economic Elites and Russian Regional Elections, 1999–2000," *Europe-Asia Studies* (2002).

Valeri V. Patsiorkovski is professor and laboratory chief at the Institute for Socio-Economic Studies of the Population, Russian Academy of Sciences, Moscow. His research interests focus on rural sociology and informal networks in rural Russia. He is the author or coauthor of numerous books and articles, including most recently *Household Capital and the Agrarian Problem in Russia* (2000).

John Reppert is executive director of Belfer Center for Science and International Affairs, Harvard University. He joined the center after retiring as a brigadier general in the U.S. Army, in which he served for thirty-three years. His research focuses on international arms control and military affairs in the former Soviet Union.

Louise I. Shelley is professor in the Department of Justice, Law, and Society in the School of International Service at American University. She is the founder and director of the Transnational Crime and Corruption Center at American University. She is the author of many articles and chapters about transnational crime and Russian corruption.

Darrell Slider is professor of government and international affairs at the University of South Florida. He has published widely on Soviet and post-Soviet politics, with a special focus on regional policy and federalism in Russia.

Nathaniel Trumbull is codirector of the Transboundary Environmental Information Agency in St. Petersburg, and a doctoral candidate in the Department of Geography at the University of Washington.

Craig ZumBrunnen is professor of geography in the Jackson School of International Studies at the University of Washington. He is also codirector of the Program on the Environment at the University of Washington. He has published extensively on Soviet and Russian environmental problems. Currently, his research focuses on the role of the Internet in solving Russian environmental problems, funded by the National Science Foundation.

Preface

Stephen K. Wegren

Even though the focus of this book is on the policy challenges faced by Russia at present and those it will face in the future, the origins of the volume date back a decade to the time when Boris Yeltsin unleashed political and economic reforms that changed the face of Russia. Just as few people foresaw the collapse of the Soviet Union and its empire, no one really understood what "shock therapy" would mean and what consequences it would spawn when Russia began its course of reforms in 1992. Most analysts agree that Russia not only is different today, but is considerably weaker and poorer, and faces more policy challenges than it did ten years ago.

While Russia may be different, many of its problems remain the same. Problems carried over from Soviet times, so many of the problems that existed in 1992 not only remained unresolved during the 1990s but actually worsened. These old problems combined with new challenges brought on by privatization, globalization, economic integration, and the breakup of the USSR. In number and complexity, the policy challenges confronting Russia today make this perhaps the most difficult time Russia has faced since 1941.

While most analysts agree on what has happened to Russia, there is sharp disagreement over the causes. Some analysts argue that Russia never applied shock therapy and that incomplete reforms explain the lack of economic growth. Others point to market reforms themselves, combined with naive and perhaps even harmful advice from international lending agencies. These factors, combined with rampant corruption and a physically weak president during much of the 1990s, help to explain the social decay and economic depression Russia experienced.

This book is not intended to solve the question of who is guilty. I personally do not accept the argument that economic collapse was inevitable during the transition. Russian policy makers were free agents who made deliberate policy decisions. Whether they understood what they were doing, whether they were blinded by the new opiate of the people (privatization), or whether they had any inkling of the consequences their actions would bring, are all interesting questions but essentially irrelevant to the task at hand. If Russia is to reemerge as a significant international actor in this century, the policy challenges it faces must not only be understood but also addressed in a positive manner. For Russia, this means significant improvement in domestic socioeconomic conditions at the same time international security issues are being managed.

The genesis of this volume is twofold. More than a decade of traveling to Russia several times a year brought personal dismay at Russia's economic and social disintegration. When conditions were especially bleak, I remember wondering to myself on many occasions how Russian society and the economy continued to function at all. What kept it going? Why did people work when they were not paid? How did people endure tremendous hardships and shortages so patiently, without resorting to violence or revolt? When I returned, I would describe to my students and to the community the realities of Russian reform. Needless to say, my listeners were amazed, curious, appalled, and captivated. Now that President Putin has brought some stability to Russia, the questions are less about whether Russia will collapse. Instead, the most pertinent issues concern to what degree Russia will successfully address the range of new and old policy challenges and, in doing so, begin to rebuild the nation, perhaps attaining new levels of development.

It was the combination of potential rebirth, but plagued with problems, with the opportunity to organize a conference on Russian security and policy challenges that was the second key factor in the genesis of this book. The conference was funded by the John G. Tower Center for Political Studies at Southern Methodist University and was held on the campus of Southern Methodist University, in April 2002. The conference provided a wonderful opportunity to bring together a broad range of interdisciplinary experts on some of the most important policy challenges facing Russia. I was exceedingly fortunate to be able to put together such a well-known and respected group of experts. I would like to thank the contributors not only for a truly interesting and useful conference but also for their excellent contributions to this volume, which are based upon the authors' original research. The reader will find a broad range of topics in this volume, so much so that it would be virtually impossible for any one writer to master all of the different policy areas contained herein. Without the cooperation of the contributors, this book would not exist.

Therefore, the purpose of this volume is to analyze a broad range of policy challenges confronting Russia. Our collective hope is that the reader will gain from each chapter not only a sense of the origins of each particular problem, but also an understanding of the policy challenge as it exists today and the prospects for its resolution in the future.

Introduction: Policy and Security Challenges in Russia

James R. Millar and Stephen K. Wegren

Enormous changes have taken place in the former Soviet bloc over the past ten years. The transformation of the former Soviet bloc started with the Gorbachev reforms, spread to Eastern Europe where Communist regimes fell, and, as part of the larger feedback loop, culminated in the collapse of the Soviet Union itself in 1991.[1] Since then, the post-communist world has changed in fundamental and most likely, irreversible ways. These changes have brought new opportunities, as well as new challenges, to policy makers in Moscow and to nations that interact with Russia.

In the post-communist era, Russia confronts a series of challenges posed by international, political, and economic insecurities occasioned by the collapse of the Soviet empire. The approach to be used in this introduction utilizes a common concept in economics: supply and demand.[2] Our concern is especially with the enormous increase in insecurity in Russia since 1990 for the elite members of society, including regional and central governmental administrators, and for the general population at large. The challenge is that the supply of security has decreased sharply, increasing the cost of security and the level of security that can be afforded in all dimensions of life.

The price of security is the opportunity cost, that is, the other personal and/or societal objectives that must be sacrificed to attain a given level of security. The sharp upward shift in the supply schedule means that a large portion of previously satisfied demand for security can no longer become effective demand. The loss of security has been occasioned in part by the loss of the Soviet empire, the breakup of the USSR, the loss of administrative continuity, mistakes in economic reform strategies, and the consequent sharp decline in real GDP from 1989 through 1999. Actions of foreign governments, especially the United States, have also contributed to the decrease in the supply of security for Russia.

The large gap between the current demand for security and its diminished supply in each dimension of Russian social, economic, and political life should help us understand the fear, anger, and humiliation that most Russian citizens feel about the loss of empire and the economic decline of Russia. It may also help us to understand President Putin's continued popularity, for he has been successful in providing at least the perception of an increased supply of security for society at large.

This volume is organized into three sections, each of which analyzes a type of policy challenge faced by Russia and its leaders: international security challenges, sociopolitical challenges, and socioeconomic challenges. In each of these three categories, Russia faces significant policy challenges in resolving the outstanding problems, which is to say, none of these policy challenges is likely to be resolved quickly or easily. Over the longer term, the extent to which Russian policy makers are successful not only will go far in determining what kind of country Russia becomes, how Russians will live, and Russia's future place in the world order, but it will also influence mass and elite perceptions of security. Using the concept of supply and demand and applying it to perceptions of security, the remainder of this introduction surveys the policy and security challenges confronting Russia today and in the foreseeable future. The review below is thematic, and references to authors' chapters do not reflect the order in which they appear in the book.

The Decrease in the Supply of Security: The Population at Large

Every society faces an inescapable trade-off between equity and efficiency in the provision of security to its members. Efficiency requires that individuals be compensated fully for their contribution to the social and economic system. Otherwise, their incentive to contribute (efficiency) will be diminished, and the total social product will decline. On the other hand, not every member of a society is able to contribute to the social product. The very young and the very old, as well as the ill and challenged members of society, are not able to contribute. Equity, defined as fairness, and humanity dictate that those who have contributed in the past and those who are expected to contribute in the future, along with those physically or mentally unable to do so, be supported by those who can. Support in this context includes medical care, safety in one's home and person, and so forth, as well as the provision of a minimum standard of economic existence. The elderly and the young are supported according to the great principal of reciprocity among generations. The physically and mentally challenged must be supported out of a sense of humanity. In either case, those who can and do work are taxed to support those who cannot.

If equity dominates in a society, efficiency and the product to be distributed diminish. If efficiency dominates, equity suffers, and many citizens must do without a minimum standard of living. Each society must face this trade-off, and solutions differ. The Soviet Union followed a Marxist ideology in which equity dominated efficiency. The United States, as the market economy par excellence, sets the trade-off much more in favor of efficiency. The transformation of Russia from a socialist state to a market economy has necessarily involved a sharp increase in the social value of efficiency at the expense of equity.[3]

An example of the trade-off may be helpful. Before shock therapy, most of the highly desirable consumer goods and services were underpriced, which meant that queuing was necessary. From an economic standpoint, acquiring a fixed price good in short supply requires two budgets: a money budget and a time budget. Those citizens who had more time than money benefited from queues. Those with more money than time had less effective purchasing power than would be the case in a market economy where price is the sole allocation criterion. When prices were liberalized, the prices of these goods rose relatively, and queuing was no longer necessary. As expressed by the ordinary citizen, goods were in the stores, but many could not now afford them. This represented a decrease in economic security for the less well-off members of society. Access to medical care, educational, and employment opportunities have similar opportunity costs.

Consider the kind of security ordinary Soviet citizens had before 1989:

1. Full employment, complete job security, and a lifetime pension (which was adequate to cover living costs);
2. Universal health coverage;
3. Protection from inflation;
4. De facto lifetime claims to secure housing;
5. Effective police protection;
6. Subsidized utilities and transportation; and
7. Universal, free education.

The "Soviet Interview Project," which interviewed thousands of immigrants to the United States in the late 1970s and early 1980s, revealed that these former Soviet citizens not only were conscious of the above benefits of Soviet life, they were also aware that they paid a price for them in terms of the economic efficiency of the Soviet system.[4] These benefits were provided to everyone, and they were underpriced. Consequently, most of them required queuing and were not allocated efficiently. They represented an attempt to put a floor under the weakest citizen, but the trade-off meant, of course, that a ceiling was also imposed on the strongest. A majority of current Russian citizens still regard those years with nostalgia. A recent public opinion survey revealed that more than 55 percent of the population would like to return to its pre-reform standard of living.[5]

The "Soviet Interview Project" and subsequent surveys in Russia reveal that the population at large did not and still does not support private ownership of heavy industry, defense industries, banking, transportation, communications, and utilities. The principal reason appears to be a sense of inequity. No one individual should own such large, important property, upon which everyone depends for life and security. These concerns should be collectively owned. When Gorbachev decided against Shatalin's 500-day plan, he described it as "irresponsible" because it called for the abolition of central planning, the introduction of markets, and the privatization of industry and land.[6] His view was shared by the majority of Russians at the time—and ten years later a significant percentage still hold that view.

Thus, the collapse of the Soviet social system did not entail simply an enormous decline in GDP and GDP per capita; it undermined a system of social claims and rights. Soviet citizens also lost bragging rights as members of a superpower, and in that sense suffered a loss of national pride and dignity. In the process, the supply of economic, social, and societal security was severely diminished. Restoration of a sense of security—not to mention the reality of security—represents an enormous challenge to Russian leadership.

In Russia today, a range of domestic social and economic issues threatens the security of the population at large and confronts policy makers with significant challenges. First, employment protection and job security are no longer assured. Although official data most likely understate the true extent of the problem, unemployment at the end of the 1990s was over 13 percent among the "economically active" population. Among the most highly affected were the young (aged 18–29) who had completed a general secondary education and were married.[7] Even when a person was employed, wage arrears and the erosion of standards of income due to inflation were constant problems. Curiously, despite deteriorating employment, Russian labor was for the most part quiescent during the 1990s. Debra Javeline, in her chapter, argues that quiescence not only is politically important but also has significant health and psychological effects on the labor force.

But worker quiescence is not the only threat to personal security. First, there is the rising threat of crime, which in several regions has become a significant problem. Especially in large cities, residents no longer feel safe, as is shown by the proliferation of steel doors installed on personal apartments, security systems for cars and apartments, and even personal guards for the wealthy. In Moscow, there even have been reports of pension-age

individuals being murdered for their apartments. Louise Shelley explores the
significance of crime in Russia today, analyzing regional patterns and types
of crimes that are most prevalent.

Beyond the rise of crime as an everyday fact of Russian life, the deterioration of health care, the birth to death ratio, the health of the young, and the
general health of the nation confront policy makers with perhaps their greatest challenge. Some analysts estimate that the Russian population could be
one-half its 2000 level by mid-century. Others are not so pessimistic, but
virtually everyone agrees that the contraction of the Russian population will
continue for at least another ten to twenty years.[8] The stark fact of the matter
is that more Russians are dying each year than are being born, as reflected in
the population coefficient, which dropped from +2.2 in 1990 to −6.7 in 2000
(for the entire population).[9] Moreover, fertility rates are dropping, infant mortality has increased, life expectancies have declined, especially for men, and
the incidence of respiratory and cardiovascular diseases has increased. Demographic trends are explored in the chapter by Timothy Heleniak, while
some of Russia's environmental problems and their effects are examined by
Craig ZumBrunnen and Nathaniel Trumbull in their chapter.

Finally, during the 1990s a "food security" problem arose for individuals
and the nation at large. For the nation at large, the problem was the increase
in food supplies coming from foreign imports, particularly meat products,
and especially in large cities, for which it is estimated that 70 percent of meat
supplies come from imports. For individuals, the problem consisted of decreased per capita consumption, lower caloric intake, and less variety in the
daily diet. After 1992, the typical Russian found that food, which was so
cheap during the Soviet era (as a consequence of pricing policies that led to
excess demand), no longer was protected by state pricing policies and did
not benefit from consumer subsidies. Thus, food became much more expensive, accounting for one-half or more of the family budget. As a consequence,
Russians changed the structure of their diet and consumed fewer calories,
particularly from proteins. The challenges of food security and rural revival
are analyzed in the chapter by Wegren, Belen'kiy, and Patsiorkovski.

The Decrease in the Supply of Security: The Political and Military Elite

As any observer of Russia is aware, Russia has become poorer and more
highly stratified.[10] The gap between the very rich and the very poor grew
significantly during the 1990s. Likewise, it is possible to discern a stratifica-
tion among elites. There is considerable evidence to suggest that at least
some economic elites have benefited from reform and have improved their

economic status and condition during the past ten years. Thus, for many of the economic elite, there has been an increase in the supply of security. For the noneconomic elite (political and military elites), the loss of security is a bit more complex. The noneconomic elite are affected by some of the same aspects of decreased security as the population at large (crime, health, and environmental deterioration). Beyond that, however, elite members of the former Soviet Union have experienced other elements of insecurity, namely to their prestige and to their position (broadly defined). Regarding prestige, one of the most dominant facts of life has been the loss of empire and its trappings:

1. Loss of the post–World War II empire that had been won at great cost;
2. Loss of the pre-Bolshevik empire in the breakup of the USSR;
3. Decline in military power and in the ability to project power beyond national borders;
4. Loss of status as a "superpower";
5. Expansion of NATO and collapse of the Warsaw Pact;
6. Repudiation of Russian influence in East-Central Europe and a decline in influence in the former republics of the USSR;
7. Presence of U.S. and other non-Russian troops in Afghanistan and Central Asia;
8. Loss of an ideological rationale for the empire;
9. Loss of cultural influence within the boundaries of the old empire;
10. The end of an ideological vision; ideological drift.

There are other factors that could be listed, but it should be clear that former members of the Soviet elite have suffered a very large decrease in the supply of security that the USSR's empire status conferred. The diplomatic community has sought to diminish the loss of global status by opposing NATO expansion and by attempting to create a counterweight to the perceived U.S. drive for world hegemony through alliances with some of the Soviet Union's old allies such as Iran, Iraq, and even China. The Russian military has not been able to face the reality of its loss of status and military standing and reform itself in accordance with its new, very limited prospects. John Reppert in his chapter examines national threat perceptions as expressed by political and military elites.

A second type of insecurity concerns specifically the political elite and their positions, and here we have two things in mind. First, there is an ongoing occupational and generational turnover among members of the elite. Studies have shown less of a recirculation of Soviet-era nomenklatura and more of an "ascendant elite," thereby replacing the old political class.[11] A second aspect is analyzed by Darrell Slider, Christopher Marsh, and Joel Moses in their chapters on center-periphery relations, the development of civil society, and Russia's struggle with democratization, respectively. This second aspect of elite insecurity concerns the rights, influence, and power that some members of the political elite have vis-à-vis other members of the political elite. For Slider, the questions concern what regional governors can do (or cannot do), how much freedom of action they have, how much influence they wield, and the balance of power with the center. For Marsh and Moses, the critical issues are: How far has civil society developed, that is, how much freedom does the elite have to act and speak without fear of reprisal? How deeply ingrained is a sense of civil liberties, outside the bounds of state interference? What impact do political institutions have on democratization, and how do they operate? Of course, these latter issues affect not only the political elite, but society at large as well. Nonetheless, members of the political elite are even more sensitive to these issues because of their elevated position in society.

The Decrease in the Supply of National Security

During the post-Stalin era, the Soviet Union was secure within its borders and was a power to reckon with prior to 1989. National security has declined precipitously since then, and there is little likelihood that things will turn around. Russia has become a third-world country, with all the economic, social, and political uncertainty that condition entails. The nation and its political elite have been humiliated by becoming subject to lectures and restrictions from the International Monetary Fund (IMF) and other international financial and political organizations. As the reality of decline has made clear, Russia was not as secure as was previously assumed by friends and foes alike, but in matters of security, perception is reality.

In the contemporary era, the perceptions in Moscow are that Russia faces a twin threat of being marginalized internationally and of having its domestic weaknesses exploited by other actors. This twin threat encompasses a number of foreign policy dimensions, including: (1) perceived linkages between NATO enlargement and "energy security," since most of Russia's oil exporting terminals are located in former Soviet republics and Soviet bloc states that are prospective NATO members (e.g., the Baltic states and the perceived threat of "Gulf War II on the shores of the Caspian Sea"); (2) the "Russian diaspora" problem in the "near abroad" (former Soviet republics) in the absence of a coherent strategy and commitment of resources to deal with this issue; and (3) perceived linkages between globalization and militant Islamic movements along Russia's periphery and the threat of Muslim-backed terrorism threatening Russian interests and the Russian homeland. Mikhail Alexseev analyzes the security challenges, Russian perceptions, and Russian policy toward the near abroad. Gregory Gleason, on the other hand, looks at nationality challenges both within Russia and on its borders. He is concerned first with the prospect of breakaway republics situated inside Russia, similar to Chechnya, as well as with broader security challenges along the north-south axis, in which Russia shares the longest contiguous border in the world with Kazakhstan. Russia also has security interests and commitments throughout Central Asia (e.g., troops in Tajikistan), as well as in Armenia, Azerbaijan, and Georgia.

In addition to the "near abroad," Russia has other security challenges along its borders, in particular in Asia. Asia represents one of Russia's largest foreign policy challenges, and Dale Herspring in his chapter analyzes Russian-Asian relations. Although Asian relations are becoming increasingly important to Russian foreign policy as a whole, Russian capabilities in the Russian Far East have weakened militarily during the 1990s; and economically the region remains underdeveloped. With regard to China, Russia's Far East is experiencing migratory outflows westward. At the same time, the number of Chinese in the Far Eastern region and along the joint border exceeds 80 million, leading to Russian fears of "border creep." Chinese workers comprise ever-increasing percentages of workers and illegal immigrants. Both Russia and China are deeply committed to retaining a military balance in the region, a need arising from decades of mistrust. At the same time, there are mutual interests. Russia has a rising economic interest in trade with China, and benefits from the sale of military weaponry. Russia and China have a mutual interest in preventing the rise of the United States as a world "hegemon." Both sides are deeply suspicious of America's missile defense plans. Concerning Japan, relations remain stagnant due to continuing disputes over territory. The policy challenge is to break the long-standing impasse so that relations can improve, thus opening the prospect of large-scale Japanese investment in Russia's Far East.

U.S.-Russian Relations and the U.S. Monopoly on Global Security

U.S.-Russian relations have experienced numerous ups and downs since the breakup of the USSR. The nature of the U.S.-Russian relationship, thematically and chronologically, is analyzed by Herbert Ellison in his chapter. However, a special section of the introduction is devoted to this topic, not because U.S.-Russian relations are as central to world stability and peace as were U.S.-Soviet relations, but because, as the sole superpower in the world today, the United States in effect has a monopoly on global security. The U.S. has the wealth and the military power to reduce the supply of security to any part of the world or for any specific country. It has the power also to increase the supply. This power is not absolute, but no monopoly has absolute power.

A monopoly allows the United States to increase the cost of security, i.e., to decrease the supply, to other countries. The United States has taken actions since the collapse of the Soviet Union and its empire that it would not have attempted earlier. John Foster Dulles dreamed of "rolling back" Soviet power in Eastern Europe, but when it came to the crunch, he hesitated. That dream has now been realized. The United States is taking action in the Caucusus and Central Asia regarding access to oil and gas supplies that could not have been attempted just a decade or so ago. It can do so because Russia cannot afford to challenge America's monopoly on security. There are limits to the ability of the U.S. to ignore Russia's geopolitical and global interests, but they are not very constrictive by comparison with the days of the Cold War.

The U.S. has contributed significantly to Russia's current perception of insecurity, both domestic and global. The following events in U.S.-Russian relations have contributed to Russia's international insecurity.

1. Support for NATO expansion eastward;
2. Post-September 11 U.S. actions in Afghanistan and the positioning of troops in Central Asia and Georgia;
3. Bombing of Yugoslavia;
4. U.S. withdrawal from the ABM Treaty;
5. Support for the IMF and the Washington Consensus;
6. Extending the U.S. lead in military and civilian technology;
7. Treating Russia as a second-class power;
8. U.S. unilateral threats to attack Iraq;
9. Mini-trade wars with the U.S., which in 2002 imposed a tariff on Russian steel;
10. European Union expansion eastward, with the possible exclusion of Russia.

Thus, a little more than ten years after the breakup of the Soviet Union, the Eurasian continent has been divided at the expense of the Soviet Empire. EU expansion is drawing East-Central Europe away from Russia. NATO expansion is extending NATO's line of defense eastward toward Russia. In fact, a new economic and military curtain is forming in Europe. On one side are the states of the former Soviet Union, minus the Baltic States. On the other side are central and eastern Europe, possibly minus Romania and Bulgaria. This division was being reinforced by Putin's adoption of Primakov's policy of attempting to create a counterweight to the U.S. through alliance with the former republics and allies of the Soviet Union. This was an alliance of losers, and a costly one for Russia, because money in an alliance, like water, runs downhill.

Meanwhile, opening the Russian economy to international commercial and financial flows has proven painful and controversial. After years of autarchy and tightly controlled domestic economic processes, Russian economists have had to learn how to adapt to global forces. Some lessons, such as the 1998 financial crisis, have been extremely costly. Other lessons, such as the need for foreign direct investment, have yet to be realized. There is considerable resentment and fear over the activities, actual and prospective, of foreign enterprises and capital in Russia.

Thus, America cannot be surprised to learn that it is not popular in Russia. Public opinion polls show a general trend of mistrust of U.S. motives. No one loves a monopolist. But a monopolist finds it difficult to act any other way. The security of the superpower monopolist is enhanced by increasing the cost of security to the rest of the world. Weaker states have a choice: to merge with the monopolist, thus becoming a client state or a junior partner, or leave the field. Prior to September 11, Putin was attempting a third path by creating a countervailing alliance. Because the prospective members of that alliance were weak states, this would have amounted, in the end, to the second alternative.

Russia's Security and Policy Challenges Under Putin

In his two years in power, Putin has achieved a significant degree of increased security for Russia and the Russians. In the domestic political realm, after more than ten years of turmoil and erratic reforms, Putin has shifted the government's strategy to one of working with the State Duma, not against it. Putin and his government have adopted a strategy of co-optation, consensus, and cooperation in working with the Duma to introduce much-needed legislation. Gone are the days of overt confrontation. To be sure, underlying the policy of consensus is the implicit threat of reprisal, if not coercion. Certainly, Putin's reaction to the oligarchs who challenged him gave pause to others. In addition, control over regional powers and the orchestration of politics in the Duma that led to the resignation of Communist committee chairmen in spring 2002 send the desired signal to those who might think of exerting their freedom of action without the approval of the Kremlin.

In the domestic economic realm, the growth of the economy following the 1998 financial crisis has also increased the sense of security of the population at large. This situation is rather fragile, because it depends on factors neither Putin nor his advisors control: the devaluation of 1998, which made imports more expensive and which gave a boost to domestic producers, and the rise in the price of oil since 1996. Putin also appears to have a firm hand on the tiller, which increases the public's sense of security and Putin's popularity.

In the international realm, Putin's decision to turn to the West after September 11, if maintained, is both the most economically rational and most diplomatically promising course of action available. Since 9/11, Putin has played down the significance of NATO expansion, of the U.S. intervention in Afghanistan, U.S. withdrawal from the ABM Treaty, collateral moves into Central Asia, and the stationing of U.S. troops in Georgia. It would appear that Putin is prepared to play the role of junior partner, and, by dampening the negative reaction of the elites to these developments, he has managed to increase Russia's sense of security as well. He remains highly popular with the public at large, mainly because he is seen as a source of stability and order and thus of security. However, because many elite segments of Russian society are strongly opposed to these actions and still hope to restore Russia to a unique but comparable position in global politics, Putin's own political security may have been diminished. Time will tell.

Notes

1. The literature on Gorbachev and his reforms is quite extensive. For a very good overview of all aspects of Gorbachev's reforms, see Archie Brown, *The Gorbachev Factor* (Oxford: Oxford University Press, 1997).

2. The idea to apply supply and demand to security was originally conceived of and presented by James R. Millar in his keynote speech for the conference on "Russian Security Challenges in the Twenty-First Century," Southern Methodist University, April 6, 2002.

3. For an early analysis of this trade-off, see Ed A. Hewett, *Reforming the Soviet Economy: Equality versus Efficiency* (Washington, DC: The Brookings Institution, 1987).

4. See James R. Millar, ed., *Politics, Work, and Daily Life in the USSR: A Survey of Former Soviet Citizens* (New York: Cambridge University Press, 1987).

5. UPI, "Poll: Russians Long for Pre-Reform Era," Johnson's Russia List #6008, January 7, 2002.

6. James R. Millar, "Prospects for Reform: Is (Was) Gorbachev Really Necessary?" in J. J. Lee and Walter Korter, eds., *Europe in Transition: Political, Economic, and Security Prospects for the 1990s* (Austin: The Lyndon B. Johnson School of Public Affairs, 1991), pp. 77–78.

7. See *Trud i zaniatost' v Rossii* (Moscow: Goskomstat, 1999), pp. 118, 146, 149.

8. See *Ob osnovnykh tendentsiiakh razvitiia demograficheskoi situatsii v Rossii do 2015 goda* (Moscow: Goskomstat, 1998).

9. *Demograficheskii ezhegodnik Rossiiskoi Federatsii* (Moscow: Goskomstat, 1994), p. 41; and *Demograficheskii ezhegodnik Rossiiskoi Federatsii* (Moscow: Goskomstat, 2001), p. 61.

10. See William Moskoff, *Hard Times: Impoverishment and Protest in the Perestroika Years* (Armonk, NY: M.E. Sharpe, 1993); and Bertram Silverman and Murray Yanowitch, *New Rich, New Poor, New Russia: Winners and Losers on the Russian Road to Capitalism* (Armonk, NY: M.E. Sharpe, 1997; rev. ed. 2000).

11. David Lane and Cameron Ross, "Russian Political Elites, 1991–1995: Recruitment and Renewal," *International Politics*, vol. 34 (June 1997), pp. 169–192.

RUSSIA'S
POLICY CHALLENGES

PART ONE

INTERNATIONAL
SECURITY CHALLENGES

Russia's Threat Perceptions

John Reppert

While the internal behavior of any nation toward security issues, as well as its international actions, must be the result of a complex set of interactions and considerations, a fundamental driver in the formulation of security policy is an internal "threat assessment" calculation. The end of the Cold War has resulted in a reassessment of this consideration in most industrialized nations on both sides of the old barriers. However, nowhere has this calculus been more thorough or more drastic than in Russia. For Russia the political changes affecting security calculations changed dramatically, as they did for all nations, but simultaneously their capabilities to respond to perceived threats also changed more significantly than in any other country. This study provides an assessment of Russia's current perception of threats upon which it prioritizes its security decision-making.

When the Soviet Union's last leader, Mikhail Gorbachev, assumed his duties as general secretary in March 1985, he entered office with a long list of concerns about the internal situation within the Soviet Union, but with a fairly short and relatively comfortable "security" concern list. A strategic balance had been struck with the U.S. and NATO and, despite the introduction by President Reagan of the Strategic Defense Initiative, the Russian military-industrial complex and its vast strategic arsenal were certainly adequate to deal with issues in that category. The border with China was always a potential source of friction with a large and powerful neighbor. Here too, however, Moscow and Beijing were working on specific plans to delineate the border to their mutual satisfaction, and both sides saw a course ahead that would reduce their considerable force deployments along their shared border.

In retrospect, the most remarkable thing is the collection of issues that were not on Gorbachev's short list of security concerns. The fate of the Warsaw Treaty Organization was not an issue. The survival of the Soviet Union as a nation state was not on the worry list. Internal conflicts based on nationalism, ethnic, or religious tensions capable of draining the resources and the morale of the nation would have been difficult to imagine.

Only slightly more than five years later, General Secretary Gorbachev quietly turned out the lights on the Soviet Union on Christmas Eve 1991, and President Yeltsin inherited much of his mantle as the leader of a Russian state. Unlike his predecessor, Yeltsin had a much longer and more serious list of security concerns. Under his predecessor he had witnessed the unification of Germany within the NATO structure and the total collapse of the Warsaw Treaty Organization, and most recently had been an active participant in the demise of the Soviet Union and the creation of fifteen sovereign states. While there was a real shared belief in the promise of strategic cooperation with the U.S. and Western Europe, both of whom had celebrated the Soviet collapse and the emergence of a national commitment to democracy and a market economy, the vast Soviet nuclear arsenal was now on the territory of four sovereign nations (Russia, Ukraine, Kazakhstan, and Belarus). The Soviet Ministry of Defense, the Soviet Ministry of Internal Affairs, and the Soviet Federal Border Guards, which had served as the physical guarantors of Soviet security, were now all being disassembled into component parts matching the borders of the new nations. The military-industrial complex that had been so intricately integrated throughout the component parts of the Soviet Union now formed a bizarre patchwork quilt of component parts across the new political landscape. The collapse of the authority of the Soviet Communist Party meant that disputes with China to the south were no longer conducted on the basis of a shared ideology. Ethnic and religious differences suppressed in the Soviet system were rapidly emerging as serious threats to order and stability.

Now a decade into the life of the new sovereign Russian state, the kaleidoscope of security concerns has rotated once more to create new lists and new priorities. When President Putin gathers his National Security Council, how do they assess the most immediate and longer range concerns, and how do the resources of the state match the needs created by these concerns?

Many of the "threats" now highest on the list of the current Russian president, unlike his predecessors, are internal to his own nation. These concerns can effectively be placed into three groups: internal fracturing, economic shortfalls, and demographic decline. Looking beyond his borders, the Russian president sees dangers in NATO expansion, American hegemony, and Chinese aspirations. In a few cases, most recently terrorism, the threats operate independently of national borders and threaten from both within and without.

Internal Threat Perceptions

The current and future behavior of the Russian state is most likely to be understood if these perceptions are examined in more detail in terms of how the leadership of Russia understands the issues. These perceptions can be split into the threats themselves and the manner in which the state could potentially respond to the threats.

Internal Fracturing

As would be predicted based on Maslow's hierarchy of needs, the highest-ranking concern of President Putin for both his regime and his state is survival. While he can, at present, set aside the possibility of a mutually fatal nuclear duel with the U.S., an invasion of his nation by NATO, or even a large-scale military conflict with China, his nation is in some ways more at risk than when these issues were prominent. Russia continues, on and off, its seemingly interminable military conflict with the Chechens within Russia. The Chechen conflict has taken the lives of thousands of Russian soldiers and security personnel, as well as thousands of civilian and opposition fighters who are also Russian citizens. The physical destruction of industry and all the major cities in Chechnya is also a result of this conflict, one that has no clear end in sight.

Beyond the immediate issue, however, is the looming concern that other Chechnyas could emerge. Russia fears, sincerely and rightly, that radical Muslim movements in the southern part of Russia, where there is still a vast Muslim population, could result in further loss of Russian blood, Russian territory, and Russian prestige in the world.

After having watched the total collapse of the Warsaw Treaty Organization and then of the Soviet Union itself, the Russians focused on territorial integrity with an urgency that would have seemed dramatically disproportionate in Brezhnev's era. The mentality of Stalingrad in that critical battle of World War II was engraved in stone in the slogan "Not one step backwards." This seems to best reflect the current Russian fear that one more fault line in their territorial composition could be fatal.

Unlike for the tsars or the communist leadership that preceded the current regime, the traditional Russian pathway to security, isolation, is no longer possible. Whatever one may think of "globalization" as a concept, it is inconceivable that Russia, even if it chose to try, could make any serious headway in closing its borders to travel, to trade, or to international communications. Compounding the problem, for economic and political reasons Russia needs to maintain relatively open borders with the area of greatest vulnerability—the states to the south, many of which were fellow republics within the Soviet Union. On a more immediate and emotional level, millions of ethnic Russians who moved to the former republics of the country when the Soviet Union was unified now represent a huge diaspora in the "near abroad." No successful Russian leadership can be indifferent to their fate or close its borders to their return.

Economics and the Military

While this kind of concern would be serious in any nation-state, it happens to catch Russia at a moment of unprecedented weakness in the security sphere, a consequence of dramatically reduced financial resources. Russia's once-vaunted conventional force capability has atrophied beyond imagination. In the final assessment of Soviet Ground Forces provided by the International Institute of Strategic Studies (IISS) in 1991 before the demise of the Soviet Union, Soviet forces were assessed to have 149 active military ground divisions in the Army, and the Soviet Air Force commanded some 5,000 combat aircraft.[1] In these same categories for Russia a decade later, the strength has been assessed at 27 active ground divisions in the army, with a combined air force and air defense force of fewer than 2,000 combat aircraft.[2] Beyond the numbers provided above, it seems nearly certain that Russia is incapable of deploying an entire, integral combat division at full manning, and it is almost certain that some 50 percent of Russia's combat aircraft are not combat ready, or in some cases not even capable of flight.

Demographics and the Military

The demographic and economic changes that have taken place in Russia further contribute to the demise of the conventional force. Russian statistics report a continuing annual population loss of more than three-quarters of a million people, a trend that has been accelerating (see the chapter by Timothy Heleniak in this volume).[3] The short interpretation is that the number of draft-age males in Russia has dropped significantly. The number of ways employed by this more limited number of candidates to effectively avoid military service through means both legal and illegal has expanded considerably. Finally, because of the significant disincentives economically and the continued perception of the brutality of enlisted life in the Russian Armed Forces, the "best and brightest" are no longer found in uniform, either as officers or as enlisted personnel. In a recent survey of Russian citizens, 84 percent responded that they "would not want their son, brother, husband, or other relative to serve in the army."[4] Sadly for Russia, even those who do respond to the draft and are found fit to wear the uniform do not always choose to fulfill their service obligation. Minister of Defense Sergei Ivanov reported in 2002 that as a function of "problems with morale," some 5,000 Russian conscripts desert from the armed forces each year.[5]

Russian military sources acknowledge that hundreds of thousands of officers have left the military other than through retirement over the past decade.[6] The literal decline in the health of the Russian Armed Forces has serious and prolonged impact on their ability to undertake security measures. By all accounts, the economic limitations of the Ministry of Defense have precluded any serious effort to fill the ranks of the military with personnel working for more extended periods under contract.

Consequences of Economic and Demographic Declines

Simultaneous with the demise of the conventional forces has been a precipitous drop in the activity of the military-industrial complex. From the mid- to late 1980s it was assessed by the U.S. that Soviet inventories in tanks, armored personnel carriers, artillery tubes, helicopters, and combat aircraft were growing at the rate of several thousands of pieces of combat equipment per year.[7] While the production of weapons by Russian industry has dropped dramatically in the last decade, even more significant is the fact that the largest consumer of Russian military production is no longer Russia. The majority of weapons systems produced in Russia currently are intended for export; and, outside of fewer than ten new strategic missile systems acquired on an annual basis in the late 1990s, the acquisition of conventional equipment was far below the minimal needs of the Ministry of Defense to maintain current inventories. The low rate of receipt of new equipment and replacement parts by the Armed Forces resulted in a formal investigation launched in early 2002 by the Russian Duma to account for the $1.9 billion allocated for arms procurement and research and development in 2001.[8]

The newly created foreign military sales agency, Rosoboronexport, is charged with two critical missions. The first is to produce a steady stream of revenue from the sales of Russian weapons to foreign states. The second is to preserve some elements of the Russian military-industrial complex by finding foreign markets for output Russia is unable to buy for its own forces.

In both force formations and the production of weapons systems, the dramatic reductions are not the result of a political reallocation of resources. President Putin understands even better than his predecessor that the economic base of the nation defines and limits the support for federal programs, such as defense. By most measures the entire Russian federal budget is less than the defense expenditures of the Soviet state only years before its demise. As a measure of divergence between the two "superpowers" of the Cold War, President Bush has proposed a $48 billion supplement to the current U.S. defense budget for fiscal year 2003. That supplement exceeds the total Russian defense budget, even if one factors in a very generous benefit to the Russian accounting through purchasing power parity.

The consequences for Russian security are found not only in the theoretical outcomes of "war games," but also in the tragedies of everyday life within the armed forces. After a year of agonizing efforts to blame the sinking of the submarine *Kursk* on some form of foreign intervention, e.g., a collision with an American submarine or a collision with a mine, the Russian government has reluctantly concluded that deficiencies in one of their own "experimental" torpedoes resulted in the explosion that doomed the sub and the servicemen on board. The greatest loss of high-level officers in the Chechen campaign has come not through combat but through the crash of Russian helicopters short on maintenance and spare parts. Finally, the explosions at munitions storage areas with losses in manpower and serious environmental problems are directly attributable to budget constraints and the decline of the forces at these facilities.

Two other critical security organs with responsibility for the physical protection of the Russian state have also been severely affected by the decline in overall Russian elements of power. The Ministry of Internal Affairs has experienced reductions in personnel and resources, though less dramatically than the Ministry of Defense. Of greater concern would be the impact on the federal border guards. They lost not only personnel and financial resources, but in many cases the literal borders they were charged with guarding. The long ambiguity over integration and independence in the Commonwealth of Independent States resulted in a porousness along national borders alien to previous Soviet experience.

A ramification of these developments with direct consequences for Russian and for international security is the increased threat that the arsenals of weapons of mass destruction within Russia could not be sufficiently secured. While this has been a matter of mutual concern since the demise of the Soviet Union in 1991, the terrorist attacks on the U.S. on September 11, 2001, provided a great increase in the sense of urgency, both in the West and inside Russia. For more than a decade, the U.S. in particular has worked with Russia to deal with acknowledged deficiencies in security for nuclear weapons and weapons-grade materials in Russia. The Cooperative Threat Program, or Nunn-Lugar program, as it is often identified, focused first on consolidating the nuclear weapons of the former Soviet Union back under one custodian, Russia. Through successful negotiations and financial arrangements with Kazakhstan, Ukraine, and Belarus, all nuclear weapons were voluntarily returned to Russian control by the middle of the 1990s.

However, cooperative progress in providing adequate security to both the weapons and the weapons-grade materials within Russia has moved much more slowly. Large numbers of facilities containing either nuclear weapons or material required to create such weapons are far below minimal standards both for accounting properly for their deadly contents and for preventing their loss. A recent U.S. assessment found in the Central Intelligence Agency, *Annual Report to Congress on the Safety and Security of Russian Nuclear Facilities and Military Forces* (February 2002), contains a long and trouble-some list of the vulnerabilities of the Russian facilities and a record of the efforts known here to take advantage of those weaknesses through theft or sale of materials.

While the significance of this vulnerability has long been shared by Russia and the West, it was the revelation that the Al Qaeda organization, which planned and conducted the catastrophic terrorist attacks in the United States, was actively working to acquire nuclear weapons or weapons-grade materials in the report provided by CIA Director George Tenet to Congress[9] that provided the latest impulse to the urgency. Russia knows full well its own vulnerability to terrorism and the costs it can exact through the recent bombings of apartment buildings and subways. There is a widespread and reasonable fear in Russia that these same kinds of terrorists, normally tied to the conflict in Chechnya, would not hesitate to employ weapons of mass destruction if they fell into their hands. The shared concern in this threat was reflected both in the most recent U.S. boost in funding to support securing the Russian materials and in Russian willingness to accept greater transparency on its part in order to receive these funds.

External Threats

When the Russian leadership shifts its attention to possible external threats, there is still the list involving NATO expansion of both members and missions, the challenge that Russia sees in the current and increasing American unilateralism in international security affairs, the longer-range problem of China's growth of power, and the one issue that most thoroughly crosses the external-internal line—transnational terrorism.

NATO Expansion

The issue of NATO expansion as a threat to Russia is one that bedeviled both President Gorbachev and President Yeltsin before Putin. After more than forty years in which NATO and the Warsaw Treaty Organization stood muzzle to muzzle along the intra-German border, it is not difficult to imagine why a citizen of Russia would view NATO as a security threat, despite its long insistence that it was exclusively a defensive organization. As the Soviets formally acknowledged the termination of the Warsaw Treaty Organization, the optimists there predicted that NATO would very soon follow suit. A fair number in the West shared this belief as well. Could NATO survive in the absence of any visible military threat from the East?

As Gorbachev accepted the withdrawal of Soviet forces from Eastern Europe and the imminent reunification of Germany, he believed by most accounts that he had "a deal" that would at a minimum prevent the eastward expansion of NATO into East Germany or beyond. This belief was rapidly destroyed as Germany was unified and came fully into NATO and the members of NATO opened links with the former Eastern European allies of Russia under the Partnership for Peace, which provided the basis for NATO acceptance of new member states.

The mood in Russia swayed as the discussions of new members intensified and the invitation for membership was formally extended to Poland, the Czech Republic, and Hungary. While President Yeltsin's ambiguity on the issue avoided a direct confrontation, the majority of Russians were solidly convinced that NATO expansion was neither a friendly nor a favorable act in terms of their security and national interests.

This concern was exacerbated in 1997 when two events happened nearly simultaneously. In the year of its fiftieth anniversary as an alliance, NATO formally welcomed the three new nations of Eastern Europe and initiated its first nondefensive, out-of-area military campaign in Bosnia. While the Russians had adjusted themselves to the passage of their three former allies into NATO, they were surprised and concerned by the decision of this "defensive" organization to launch offensive operations when it was clear that none of their members had been attacked or were under threat of attack. The rules had certainly changed.

Almost immediately after the inclusion of the Poles, the Czechs, and the Hungarians into the NATO structure, discussions were begun concerning the "next round" of NATO membership expansion. Soon nine countries were lined up with application papers in hand, this time including the three Baltic States of the former Soviet Union. The Russians expressed special distress on three issues involving the Baltics. First, they had been a part of the former Soviet Union—discussions on the international recognition of the legitimacy of that status aside. Second, there remains in the Baltic States a significant Russian-speaking population, who many Russians believe have been victims of ethnic discrimination by the Baltic governments. Third, land access to the Russian territory of Kaliningrad is only possible by transiting Lithuania. As a newly independent state, Lithuania was problematic for the Russians and had been a source of some irritation and contention. Were Lithuania to become a full member of NATO, it would mean that component parts of Russia would be separated by NATO forces.

As complicated as this issue was, the NATO decision in response to the terrorist attack of September 11, 2001, to invoke the first provision of Article V that "an attack on one is considered an attack on all" in support of the American campaign in Afghanistan was even more troubling.

While this could be cast in the wide understanding of a "defensive" action more plausibly than was the case in Bosnia, the concept that the U.S. and NATO could, without UN sanction, "declare war" on any state they perceived to be harboring international terrorists had to send a chill through Russian decision-makers.

Given this history, what is it that Russia specifically perceives to be a threat in the NATO expansion of members and missions? First, it should be noted that despite the rhetoric, serious planners in Moscow do not perceive NATO expansion as the preparation for an attack on Russia. In the author's official career, it was fairly easy to convince Russian security officials that, were NATO determined to carry out a military attack against Russia, adding the Poles, the Slovaks, or the Latvians would not be part of the plan. Without disparaging the military capabilities of these new states, the truth is that by the mid-1990s when the expansion debate was most intense, NATO held overwhelming military superiority over Russia in every category of conventional warfare. Additional forces or tanks were not required for any military operation, and the effective incorporation of all of these states would require years of effort that would consume the attention and energies of NATO.

However, beyond the physical threat of attack, Russia does have what it believes to be serious security concerns. First, the lessons of Russian history make the country especially sensitive to "encirclement": U.S. forces in South Korea and Japan, NATO forces in Afghanistan, U.S. forces in Georgia, and NATO troops in the Balkans, as well as a solid and approaching wall of NATO on the West. In the spring of 2002, betting on the outcome of the Prague NATO summit in the fall is that some five to seven new members will be formally invited into NATO, including the Baltics and possibly other former Russian close allies such as Bulgaria.

Beyond history, the Russians have two other issues. First, they fear that NATO expansion will deprive them of the formerly lucrative "arms market" provided by their former allies, who will now want to shore up their relations with the West by buying NATO equipment. Russia's arms industries are heavily dependent on exports and could be adversely affected by this type of incursion into its former markets. Second, the Russians now fully recognize their relative weakness and are transitioning from their former "superpower" status to their new position as a nation, which must rely on shifting balances of power to create influence. A Europe that is solidly within the NATO camp provides few openings in which the relative power of Russia can be shifted from one side to the other to affect balances. This explains the peculiar appeal of then Prime Minister Primakov's efforts to reach out to China, India, and the states of the Middle East.

Finally, Russia remains concerned about what it sees to be the inconsistency of NATO's relationship to Russia. The special NATO-Russia Founding Act created for consultation in the mid-1990s failed badly from Russia's perspective when put to the test in Kosovo, when it felt that its legitimate concerns were not addressed and it was excluded almost entirely from the planning or decision process. New proposals on the table for "19 plus one" have not yet been fully developed or "field-tested" to determine the extent to which Russians will have either a voice or a veto on matters they categorize as direct security concerns of the Russian state. The chief of the Russian General Staff was even more pessimistic, citing continuing NATO military exercises along Russia's borders and other NATO operational plans that continue to cast Russia as the actual or potential opponent.[10]

American Unilateralism

In many ways, a more serious and long-term "threat" for the Russian state is what it and other European states have labeled as increasing U.S. unilateralism in security issues. This phenomenon has two aspects from the Russian perspective: the U.S. willingness to make decisions on security issues without achieving a broad consensus among allies or the consent of international bodies such as the UN Security Council, and the U.S.'s unique "power" which enables it to act unilaterally.

In terms of the perceived U.S. willingness to independently make security decisions affecting other nations, Russia's concern is one broadly shared across the international spectrum and rests on recent U.S. positions on such diverse issues as the Kyoto Accord in the environmental realm to withdrawal from the Anti-Ballistic Missile Accord of 1972 and refusal to ratify the Comprehensive Nuclear Test Ban Treaty. Even after the significant and risky Russian decision to fully back the U.S. in its actions in Afghanistan after 9/11, the Russians and many others have been surprised and put on the defensive by the declaration of the "axis of evil" in the U.S. president's State of the Union message.

Were this only a matter of U.S. declarations of intention, the matter for Russia would be less serious; but Russia knows that the U.S. currently has the "power" in the broadest sense to execute its plans "with or without the help of friends and allies." A recent accounting of this relative U.S. power is provided by Paul Kennedy in his book *The Rise and Fall of the Great Powers*. In a brief article in *The Financial Times*, he cites such significant indicators as the following: At $350 billion annually, the U.S. now accounts for some 40 percent of the world's defense spending; 45 percent of the world's Internet traffic takes place in the U.S.; and 75 percent of the Nobel laureates in the sciences, economics, and medicine in recent decades live and work in America.[11] Statistics like these suggest that the current preponderance of American power in the world is neither unidimensional nor likely to be short-lived. Politically, financially, and in terms of raw military power, the U.S. has separated itself from its nearest competitors in a manner without precedent in our history. In the military dimension alone, the dramatic technological difference between the U.S. and its NATO allies in the conflicts in the Balkans was seen on both sides of the Atlantic as an undisputable but disturbing trend.

As Russia looks at the vast U.S. investment in new programs such as strategic missile defense, it sees that gap widening into a chasm, which may well exist for decades ahead. The single weapon of influence left in the Russian arsenal is its significant store of nuclear warheads, unrivaled by anyone except the United States. In this light, it sees the U.S. National Missile Defense program as a conscious and deliberate effort by the U.S. to expand its unilateral authority and reduce its vulnerability to attack or threat of attack from any nation. Russia's own severely reduced financial capacity and its pressing domestic needs guarantee that the country cannot "adequately respond" as was its mode of operation during the Cold War. Compounding this dilemma, Russia believes that there is no grouping of other nations with which it could seek security arrangements that would collectively equal the U.S., much less the U.S. aligned with NATO.

The concept of parity in arms matters that served to underlie the extensive treaty regimes of the past three decades—the Intermediate Range Nuclear Treaty, the Conventional Forces Treaty in Europe, the START I and II agreements, and even the Chemical and Biological Weapons Conventions—no longer exists. The bipolar world has become multipolar, a frequently stated Russian aspiration, but the "poles" are far from equal. Those who would see the U.S. constrained in its actions and its policies in Russia and elsewhere are increasingly required to find "balance" by pitting the U.S. against the ROW (the Rest of the World)—a most unlikely alliance from a historical perspective.

Russia increasingly realizes that it must pursue its critical national security interests not through counterbalance with the West and the U.S. in particular, but through recognizing and accommodating the interests of the world's remaining superpower. This form of realpolitik has created severe tensions for President Putin within his own government, as many still have a strong preference for the world as it was.

China

Overshadowed in urgency by the imminent actions of NATO and the United States is the long-standing security concern that Russia has about the future of China (see in addition to this discussion the chapter by Dale Herspring). Three decades ago, when tensions between the two huge Asian powers arose, the then Soviet Union was able to generally deter the Chinese in their ambitions, and when deterrence failed, they could reliably defeat them in combat on any rung of the ladder of escalation. Outside of the top rung with nuclear weapons, this Russian confidence has been badly shaken by the decline of their own conventional capabilities.

This decline has been matched by recent massive increases in Chinese military spending. *The Washington Post* and *The Financial Times* report that China is expected to publicly announce an increase of 17.6 percent in its defense spending for 2002 alone, building on a 17.7 percent increase last year.[12] China's current five-year plan predicts annual rates of budgetary increase for defense spending of 15–20 percent. In 2000, China was the world's largest weapons importer, a title it almost certainly maintained in 2001. The combination of a collapse of Russian military strength and a rapid increase in Chinese military capability is obviously threatening, and this gap seems likely to increase in the future.

For common and separate reasons, Russia and China have successfully defused their tensions on the long-shared border, allowing both sides to reduce the huge conventional forces they had earlier assembled there. Beyond that, Russia finds itself in a particularly unique dilemma in serving as the largest arms supplier of sophisticated modern military weaponry to a country that some in the government and many in the Russian Far East think they will eventually have to fight. A host of senior Russian defense officials have stated their predicament quite unambiguously to the author. China is Russia's largest cash customer for modern military technology. According to *The Washington Post*, China in 2001–2002 purchased 72 SU-27 fighter-ground attack aircraft, 100 S-300 surface-to-air missiles, 10 IL-76 transport aircraft, 4 Kilo-class submarines, and 2 Sovremenny-class destroyers. China has also signed a contract to assemble at least 200 more SU-27s within China.[13]

Without that funding, several critical Russian defense industries, unable to sell their output to a poor Russian Ministry of Defense, would simply collapse. This would leave Russia without the capacity to reconstitute an effective military force, even if their financial situation improved dramatically in the years ahead. Either the Russians arm the Chinese with weapons they may one day have to contend with in a struggle with their Asian neighbor, or they allow their own defense industry to atrophy to the point that Russia would be physically incapable of providing for its own defense.

For the Russians, this is not only a matter of looking suspiciously upon Chinese intentions, it is driven by the real demographic situation along the border, where the Russia population appears to be in an irreversible decline and its neighbor to the south is equally unable to prevent growth of its own population. This demographic pressure combined with increasing Chinese affluence can only increase the pressure to the north in some way, which would territorially disadvantage Russia.

Transnational Terrorism

A final security issue that operates independently of national borders is transnational terrorism, an issue that moved from the fringe of concern in the West to become the centerpiece of our security strategy after the attacks in New York and Washington on September 11, 2001. For Russia, the September attacks were less of a revelation. For most of the last decade, Russia has labeled its conflict with the peoples of the Russian Chechen Republic as a war on criminals and terrorists. Russia's population has had its own unfortunate experience with actions attributed to terrorists as apartment buildings—even in Moscow—have been destroyed with hundreds of civilian deaths.

While there is no dispute that non-Chechens have been fighting alongside Chechen forces against the Russians and that arms and ammunition have been introduced into the conflict from the outside, the Russians previously were unable to convince most Western nations that the Chechen conflict was part of any "international conspiracy." The attack on New York and Washington by non-U.S. citizens and the revelation of the international structure of the Al Qaeda network has lent sympathy, if not full acceptance, from the international community to the Russian version of the conflict with the Muslim peoples of Chechnya. This has been further reinforced by reports from the fighting in Afghanistan in the spring of 2002 indicating that sizable numbers of the forces engaged among the Al Qaeda forces are Chechens.

As noted above, this type of activity threatens Russia in ways previously unimaginable due to the decline of its own capabilities, particularly those of the Ministry of Internal Forces and the Border Guards. For these reasons and others, President Putin has been notable in his strong backing for the U.S. campaign against terrorism and has been willing to risk his political future in supporting the war against forces he perceives as hostile and a direct threat to Russia.

Conclusion

If these are accurate assessments of Russia's own perceived threats in the near and mid-term, what are Russia's responses, which will themselves have an impact on the security of other states?

Unlike previous Communist or even tsarist leaders of his nation, President Putin has consciously avoided the temptation to pour more money into the military and weapons procurement. Despite the twin pressures of conducting an active military campaign in Chechnya and listening to the continuous pleas of his senior military officers, he has chosen a different path. By all indicators, he has made the decision that the real "power" of Russia or any nation is a derivative of its economic resources. Although he appears passionately devoted to the war in Chechnya, he has steadfastly refused to increase the proportion of funding provided to the Ministry of Defense and has concentrated instead on economic growth and reducing the national debt of Russia. On both of these counts, he has made remarkable progress in his first years in office.

It is the longer-range hope of Russia that an economic upturn will have a favorable impact on the demographic situation, both in terms of health care and in the willingness of Russian women to raise more children. These factors in turn will ease the economic and demographic factors directly affecting the sustainability of the Russian security services.

The conflict in Chechnya, however, has itself gone far less well. Despite repeated promises to reduce the military presence and turn over more authority to the Chechen administration loyal to Moscow, the president has been unable to deliver. As of this writing, casualty reports on the Russian side and claims of atrocities by the Chechens are daily activities. Therefore, the Russian fear of releasing Chechnya and its inability to conclusively deal with the conflict leave them vulnerable to concerns about future fracturing.

On the international front, where Stalin in his hour of greatest weakness chose isolation and autarky, President Putin has chosen increased integration with the world community and ties of a formal and informal nature with a wide range of nations. Even in the case of nations written off by much of the world community, e.g., North Korea or Iraq, the Russian leader has actively sought engagement as a policy and has expended Russian political capital to keep lines of dialogue open.

Russia is working intensively to reconcile differences with both NATO and the U.S. in ways that retain the maximum amount of influence for Russia.

Throughout 2001 and especially after September 11 into 2002, President Putin and his senior ministers have been exploring options to bridge U.S.-Russia and NATO-Russia differences. Some successes have been achieved, but the end game is still unclear. With China, current Russian policy is to continue the balancing act of recognizing the long-term risks, while fully exploiting the short-term opportunities for trade and international cooperation.

Russia has a set of interwoven security problems at home and abroad, which can act to compound one another. Its economic and demographic declines weaken its conventional forces while simultaneously worsening its position in Siberia versus China, for example, while the factors that led to the demographic decline impose their own costs on both the health of soldiers in the armed forces and on the potential for economic improvements necessary to pay for everything Russia needs to extract itself from its various quandaries. Thus, President Putin is faced with the traditional statesman's dilemma of reconciling means and ends even as the fundamentals necessary to Russian power are eroded by the very problems he seeks to combat.

Notes

1. *The Military Balance, 1991–1992* (London: Brassey's Publishing House, 1991), pp. 36–38.

2. *The Military Balance, 2001–2002* (London: Brassey's Publishing House, 2001), pp. 105–116.

3. *Itar-Tass*, "Russia Publishes Population Figures," February 21, 2002. Available at http://www.itar-tass.com/ru/news.asp.

4. "We Love the Army, but in a Strange Way," *Vremia Novostei*, February 22, 2002, p. 4.

5. "Russian Defense Minister Hopes Wage Hikes Will Improve Troop Morale," reported from *Izvestia* in Radio Free Europe/Radio Liberty, vol. 3, no. 9, March 8, 2002.

6. "Russian Army Suffers from Mass Exodus of Officers," *Kommersant*, February 11, 2002, p. 2.

7. *Soviet Military Power* (Washington: U.S. Government Printing Office, 1987), p. 71.

8. "Duma Seeks Probe into Army Spending," *Moscow Times*, February 19, 2002, p. 3.

9. "U.S. Certifies Theft of Russian Nuclear Material Has Occurred," *Agence France Presse*, February 23, 2002.

10. "Russia-NATO Talks Hit Snag," *Moskovskaia Pravda*, March 5, 2002, as reported in Radio Free Europe/Radio Liberty, vol. 3, no. 9, March 8, 2002.

11. "The Eagle Has Landed," *The Financial Times*, February 1, 2002, p. 1.

12. "China Plans Big Rise in Military Spending to Modernise Army," *The Financial Times*, March 5, 2002, p. 13.

13. "China Raises Defense Budget Again," *The Washington Post*, March 5, 2002, p. A10.

For Further Reading

Gulner Aybet, *A European Security Architecture after the Cold War: Questions of Legitimacy* (New York: St. Martin's Press, 2000).

Robert Legvold, "Russia's Unformed Foreign Policy," *Foreign Affairs*, vol. 80, no. 5 (September–October 2001).

Olga Oliker, *Russia's Chechen Wars, 1994–2000* (Santa Monica, CA: Rand Corporation, 2001).

Rose Gottemoeller, "Arms Control in the New Era," *The Washington Quarterly*, vol. 25, no. 2 (Spring 2002).

2

Russian-Asian Relations

Dale R. Herspring

"If we don't make a real effort to develop Russia's Far East, then in the next few decades, the Russian population will be speaking mainly Japanese, Chinese, and Korean."—President Vladimir Putin, August 4, 2000

There are few cases where Russian foreign and domestic policy are more intertwined than in Moscow's policy toward Asia, and especially toward its key neighbors: China and Japan. The fall of Communism destroyed the country's military-industrial complex (a good part of which was located in the Russian Far East, or RFE), and the end of Moscow's subsidies to the region isolated it from the rest of the country. Weapons sales to India and China—in spite of some opposition to the latter by the Russian military—are going strong and are being used, in part, to keep the country's military-industrial complex alive while Putin tries to reform it. Moscow's primary focus has been on expanding ties with China and Japan.

During the Yeltsin era, little was done to deal directly with these problems, with the exception of weapons sales to China and India. Since he has taken office, however, President Vladimir Putin has tried to deal directly with the RFE's problems. While relations with the West, and especially the U.S., have taken on a higher priority in the aftermath of the events of September 11, Putin continues to believe that Russia is both a European and an Asian power, and that the latter, in particular, is critical when it comes to solving the RFE's problems.

Post-communist Russia has been fully aware of the importance of the Asian theater. In 1993, Moscow's official government foreign policy document placed the Asia-Pacific region in sixth place behind the CIS, arms control and international security, economic reform, the United States, and Europe. In March 1996, former Russian Foreign Minister Yevgenii Primakov raised the Asia-Pacific region to third place behind the CIS and East Europe.[1] Thus, Russia's

Asian policy has a dual purpose: (1) to use its ties with Asia to help rebuild its international standing in the world; and (2) to improve relations with the major Asian powers as a means of keeping the RFE alive and healthy at a time of population loss and economic destitution in that region.

Background

Before discussing Russia's policy toward Asia, it would be useful to review some facts about the Russian Far East, since it often is neglected as a region for analysis. The Russian Far East is composed of ten administrative units.[2] It extends from the Pacific Ocean to the Arctic and includes more than one-third of the Russian Federation's territory. The region is losing population—down from 8 million in 1991 to 5.2 million in 2001.[3] To make matters worse, the first to leave the region were skilled workers who could easily find employment elsewhere in Russia. In fact, many had initially been lured to the RFE with Soviet promises of temporary work with higher wages. But then the Soviet Union collapsed. Now those highly paid positions are scarce, while problems of adequate housing, electricity, heating, hospitals, and consumer goods are plentiful. Those left in the region were stranded, and lukewarm efforts to resettle the area with Russians from other areas have failed.[4]

On the other side of the border, the situation in China is quite different. There are more than 132 million Chinese living in the northwest part of the country, and that number is growing by a million persons a year.[5] If one takes only the Primorskii krai in which Valdivostok is located, the 2.2 million inhabitants are outnumbered by 70 million Chinese living on the other side of the border.[6]

There are also Chinese and Koreans living in Russia. By some estimates, there are a total of 200,000–400,000 in Russia itself, with 35,000 Chinese and 30,000 North Koreans living among the 2.2 million in the Primorskii krai.[7] Russians look south and all they see are Chinese. According to an Interfax report, "more than half of the population believe China's presence in the Far East is dangerous."[8] Siberians fear that over time the region will become Chinese by means of territorial creep.

The reality, however, is that Russians need the Chinese to supply the labor force in the Far East. One recent study, for example, argued that "Russia's population of 145 million will shrink to an estimated 126 million by 2015 if current trends continue. Unless the sparsely peopled country finds willing hands to work in its factories and farms, it faces decline."[9] The author further argues that Russia should admit 10–20 million Chinese by 2050 if it hopes to keep the region economically viable.

Economically, the Russian Far East is rich in natural resources, including fish, timber, diamonds, gold, and most importantly in recent years the large oil find off Sakhalin Island. With all of these riches, one would think that it would not be very difficult to make a profit. Unfortunately, that is not the case. Historically, the region received subsidies from the central government, and those living there also received better benefits and wages. Then came the collapse of the USSR, and, when it happened, the region's economic infra-structure deteriorated. As a consequence, the region is full of dilapidated buildings, and corruption and waste are rampant (as in other regions as well).

The economic downturn in the region started in 1989 when Gorbachev began to cut back on the armed forces. The military reduction meant fewer orders, and factories began to have problems. In fact, by 1992, "these reductions averaged about 50 percent, and affected almost all sectors.[10] Gorbachev believed that by integrating the region into the Asian market, the economic dislocation would be minimal.

When former President Boris Yeltsin took power, subsidies to the Far East were halted. When it came to transportation, for example, the cost of shipping items to Moscow went up so much that the region was cut off from the rest of the country. The same was true of energy, something that was of critical importance in an area that has such an extreme climate. As far as things like wages were concerned, not only were citizens of the RFE faced with higher prices for critical commodities, their purchasing power went down. Shipyards in the region relied on government contracts to repair surface ships and submarines. While there were plenty of ships to work on, the Russian navy did not have any money. Converting such industries to work on civilian ships would be difficult, and these shipyards would have to compete with shipyards in Taiwan, Korea, and China. The latter were not only cheaper, the quality was much higher. In essence, the result of these changes was to create what one writer called a "Chinese Wall" between the Eastern and Western parts of Russia.[11] Unemployment rose significantly as industrial output in the RFE declined about 58 percent from 1991 to 1995, as almost all of those working in the industrial military complex were laid off.[12]

During his 1996 presidential election campaign, Yeltsin announced a long-term development plan for the Far East. It called for 371.4 trillion rubles of financial support during 1996–2000. The main idea was to link Primorskii krai and Zabaikalye krai with the Northeast Asian market. Such a move would take advantage of the region's ideal geographical location. Bordering China and Korea, and located near Japan, the intent was to make this region into a showcase for Russian cooperation with the countries of Asia. In reality, however, these promises turned out to be empty, and the RFE received very little of what it was promised.

By the time Vladimir Putin assumed the presidency in January 2000, the situation in the RFE was quite different from what it had been ten years previously. Before the USSR collapsed, the Far East had sent 75 percent of its goods to the European part of Russia; but by the end of the 1990s, output was only a quarter of what it had been, and of that amount, 75 percent stayed in the region.[13] Furthermore, the region's leaders, including many in the economic sphere, were still heavily influenced by the Soviet way of thinking.

Many Russians are not overly sympathetic vis-à-vis the RFE. Many consider it a hotbed of separatism. After all, this was the area that declared its independence in the 1920s and which had often talked of going its own way. Many Russians also believed the Far East to be dominated by organized crime, run by demagogues, and full of people who were xenophobic toward the Chinese. Besides, as the logic went, if Far Easterners really wanted to stop the Chinese from infiltrating the region, they would stay there and not move to other parts of Russia. At the same time, Moscow decided that the area was critical to the country's future and realized that the region's most economically viable industry was the military-industrial complex. Very few of the factories were able to develop dual-technology products.

It is important to keep in mind that it was only recently that the process of capital movement from European Russia to the Far East started to gain momentum.[14] The most obvious manifestation of this new interest in the region was Moscow's decision to make a conscious effort to rebuild the region's military industrial complex.

For the military-industrial corporation enterprises, obtaining well-paid state orders to produce armaments for foreign countries is still the best and easiest way to survive in Russia's new economic environment. However, Russian arms exports in 1997 and 1998 dropped to $2.5 billion and $2 billion respectively, the country's worst performance since 1993.[15]

The situation had improved somewhat by 2000—increasing to about $2.8 billion. The prognosis for 2002 was much better: more than $4 billion.[16] Needless to say, the better Russia does in exporting weapons, the better off the citizens of the RFE will be.

There were a lot of factors working against the success of this program, however. For example, there were fewer problems and more money in European Russia, and that is why the ministries in Moscow devoted more of their time to that part of the country. As a consequence, "The federal agencies in Khabarovsk are often starved for funds and must rely on aid from the governor."[17]

The Russian Military and the Far East

Moscow's position in the Far East was always built on military strength. When Gorbachev began to draw down the military, few believed that the situation would deteriorate to the state in which the Russian fleet now finds itself. For example, during 1990–1999, the total number of submarines fell by 75 percent, and the total number of surface ships fell by 47 percent.[18] And it was not only the navy that suffered from the withdrawal of Russian forces.

By 1992, the withdrawal of Russian troops from Mongolia, which Gorbachev had initiated in 1987, had been completed. By 1996, the number of divisions in the Far Eastern Strategic theater had decreased from 57 to 23, the number of tanks fell from 14,900 to 10,068, the number of surface to surface missiles went from 363 to 102, and the number of attack helicopters fell from 1,000 to 310. The number of combat aircraft also decreased from 1,125 to 425.[19]

In addition to the reduction in planes, ships, and tanks, the Far Eastern Military District has also been affected by organizational problems. Discipline has been a major problem, as has corruption. Soldiers have even shot one another. Some 10,000 antiaircraft shells were washed away on one occasion, while on another Russian sailors starved to death on a remote island garrison. The military has not even had enough money to feed and house its soldiers and sailors. For many years Primorskii krai was expected to provide the armed forces with things like electricity, fuel, water, food, etc. It is no longer in a position to carry that load; indeed, as noted above, it has not even been able to supply those items for itself.

In fact, not only has the Russian military fallen apart as an institution, in the last ten years in the RFE it has become an embarrassment, as is evident from the many nuclear submarines docked in the Far East, rusting away while waiting for the day when they are dismantled. There have even been reports of incidents where individuals were able to get on board nuclear submarines that were left unguarded because there were not enough sailors to do the job. The situation improved somewhat in 1999 when the Zvezda shipyard administration agreed to cooperate with the United States and Japan in making sure that nuclear waste was properly disposed of and that submarines were dismantled in a safe manner. The bottom line is very simple: Given the disastrous economic situation, Russia's ability to project power in the region is minimal, both now and for the immediate future.[20]

Russian-Asian Relations: The Gorbachev and Yeltsin Years

This section will survey relations with two main Asian powers, Japan and China, starting with Japan. The purpose is to examine the outstanding issues of dispute, to identify areas of cooperation, to assess how relations have evolved, and to consider the prospects for the future.

Japan

In the immediate aftermath of the collapse of Communism, Russian-Japanese relations were more or less static. The problem was the Kuril Islands, a small group of four islands (some 4,500 square kilometers) north of Japan that Russia seized after the end of World War II as a result of wartime agreements between Stalin and Roosevelt. To the Japanese, these were Japanese islands, with their rich fishing grounds. However, when the Russians catch Japanese fishing boats in the area, they seize them.

In the 1956 Moscow Declaration, the Russians agreed to hand over two of the four islands. However, in 1960 Khrushchev withdrew the promise as a protest against renewal of the U.S.-Japan Security Treaty. Former President Mikhail Gorbachev understood that resolution of this problem was critical if he expected Japanese investment in the Far East. And he believed that he had to have Japanese capital and technology if he hoped to make perestroika work in the Russian Far East. Some limited progress was made. In 1986, for example, visa-free visits by Japanese citizens to the islands were permitted. And then in 1989 travel restrictions (in effect for military/security reasons) were eased. A Soviet-Japanese working group was set up in 1988 to resolve the dispute. The problem was that the Japanese would only agree to the return of all four islands, although they were flexible on how they were returned, e.g., they agreed to accept the return of Habomai and Shikotan immediately, while putting off the return of Iturup and Kunashir until later, provided the Russians accepted Japanese sovereignty over them. The Russians would not budge; the most the Kremlin was prepared to do was to give back two of them.

In April 1991, Gorbachev visited Japan, and the Joint Communiqué expressly mentioned the islands of Habomai, Shikotan, Iturup, and Kunashir and called for the resolution of the "territorial issue."[21] Given political realities, however, the bottom line was that Gorbachev could not pull off such an agreement. And the problem was not foreign policy. Rather it was internal Russian politics. In short, he was too weak to hand over to the Japanese land that the Soviet elites believed rightfully belonged to Russia.

The issue was revisited under Yeltsin. During 1992, the Japanese pressed hard for the return of the islands. In 1992, Yeltsin was scheduled to make an official visit to Tokyo. At the last minute, however, the trip was canceled "because Yeltsin could not address the territorial dispute in a way that was simultaneously defensible at home and acceptable to the Japanese."[22] The next year he apparently felt confident enough to agree, in the 1993 Tokyo Declaration, to "work to resolve the territorial issue on the basis of legality and fairness, which implies an eventual change in the status quo."[23] Furthermore, at this time Yeltsin informed his host that all Russian troops would be withdrawn from the Northern territories. From the Japanese standpoint, this agreement would become the "cornerstone" of bilateral relations.[24]

In 1997, Yeltsin and Japanese Prime Minister Hashimoto Ryutaro agreed at a meeting at Krasnoyarsk that they would work to complete a peace treaty by the year 2000. In addition, at the June 1997 Group of Seven meeting in Denver, Yeltsin and Hashimoto again discussed investment in the Far East, and Yeltsin agreed to support Japan's bid to become a permanent member of the UN Security Council. In April 1998, the two met in the Japanese town of Kawana where Hashimoto proposed a Hong Kong–type solution, in that it would require Russian acceptance of Japanese sovereignty over the four islands but delay their return to some point in the future.

The two met again in Moscow in November 1998 where they again expressed their intention to conclude a peace treaty by the year 2000. They set up two subcommittees, the first of which was to work on border demarcation, the second on joint economic activities on the islands. The committees met for the first time in January 1999.[25] As a result, there has been some progress in bilateral relations. In August 1999, a hot line was set up between the two countries,[26] and subsequent talks were held between the two militaries in Tokyo in October 2001. In September 2000 the two sides adopted a series of cooperative agreements in the economic sphere.[27]

Closely associated with the question of the Kuril Islands is the issue of Japanese investment. There has long been an expectation in the RFE that a partnership between Japanese capital and Russian labor would make the region prosperous. The problems, however, were both political and economic. Who would advise anyone to invest in a country that seemed for many years to be on the brink of total collapse? Furthermore, it looked to outsiders like it was run by corrupt officials, who in many cases would qualify as criminals in the West. There were no laws to protect investments, and even if they did exist, they were generally ignored. Japanese-Russian trade has fluctuated over the years. It was $3.48 billion in 1992, rose to $5.9 billion in 1995, then slipped to $5.38 billion in 1996, and dropped to $3.88 billion in 1997.[28] Russia accounts for less than one percent of Japan's trade. Almost all of Russia's exports to Japan are raw materials: timber, fish, coal, and metals.

China

Despite Japan's obvious importance, it is China that has continued to be the main focus of Russian policy in the Asian-Pacific region. Indeed, one could argue that Russo-Chinese relations went through three stages during the 1990s.[29] The first stage began in December 1992, when the two governments announced that they were "friendly" toward each other. For his part, Yeltsin began to focus on closer relations with China for a number of reasons. They helped balance Moscow's relations with the West; for example, he could play China off against the U.S. Such a policy would also be a means of getting China more closely involved with the RFE. The second stage began in September 1994, when the two sides decided that the relationship should best be defined as a "constructive partnership." The purpose was to give long-term stability to the relationship. The third and last stage began in 1996 when the two countries announced that they were developing a "strategic partnership." During the period from 1992 to 1999, Chinese and Russian leaders held seven summits, marking a considerable increase in Chinese-Russian diplomatic interactions.

In many ways, the 1997 summit in Moscow was one of the most important under Yeltsin, because the two countries came out against "hegemony" in any form—a code word for the U.S. In essence, Yeltsin had decided to formalize his reliance on China as a counterweight to Washington. A lot of Russians, however, were concerned about Yeltsin's cozying up to the Chinese. They believed that it was in Moscow's interest to maintain friendly ties with both China and the U.S. and Europe.[30] This debate would continue to split Russian foreign policy officials until the cataclysmic events of September 11, after which Putin would begin to focus more and more on the West.

China and Russian Weapons

For most in the West, the aspect of Sino-Russian relations that has gained the most attention has been the willingness of Moscow to provide Beijing with modern weapons, a situation that many believe will come back to haunt the Russians some day. Indeed, it is important to keep in mind that the relationship has been marred by deep suspicion.

From Moscow's standpoint, the problem is rather simple, as Evgeniy Bazhanov noted: "Moscow, desperate for capital and worried about its troubled-industrial complex, is eager to shower the Chinese with airplanes, tanks, ships and guns."[31]

The Sino-Russian arms relationship began to take on a serious form just prior to Yeltsin's visit to Beijing in December 1992. Senior Russian and Chinese military officials began annual meetings on military-technical cooperation, and they signed a protocol to formalize their long-term commitment.[32] Yeltsin signed a Memorandum of Understanding on military cooperation during his December 1992 visit—a memorandum that called for the sale of twenty-six SU-27 fighters and jet engines as well as the training of Chinese pilots. The next year, the two sides signed another protocol and agreed to the sale of more weapons. By 1994, China was Russia's number one client for military items. In July 1994, "the Chinese State Council allotted $5 billion for the acquisition of Russian military equipment."[33] Then in November 1996, Moscow and Beijing signed a bilateral defense cooperation pact. This led to the sale of two Sovremenny-class guided missile destroyers, air-defense systems, two Type-636 Kilo-class diesel submarines, large quantities of fighter planes, and air-defense systems, among other items.[34]

Since then, the two have drawn increasingly close in this critical area—especially after 1998 when Russian Defense Minister General Igor Sergeev committed Russia to help develop China's high-precision weapons system and to transfer more production licenses to China. This means China's comprehensive plan of building up conventional weapons, nuclear weapons, and delivery systems, and a massive investment in the tools and technology missile defense, information warfare, and space war, are directed mainly against America or its allies.[35] In fact, in 1998, for example, Chinese orders accounted for 25 percent of Russian arms sales, which totaled $8.4 billion.[36] Areas such as Khabarovsk krai depend on Chinese purchases of Sukhoi aircraft that are built in the area. Current plans call for China to have 200 Su-27s by 2012, although they will be coproduced, so it is not certain just how much the RFE will benefit economically.[37] In January 2002, it was announced that Russia will build two additional Sovremenny-class destroyers for China.[38]

What a lot of Russian officers find so upsetting is not only that Moscow is selling the very latest in weapons technology to the Chinese—with all that portends for the future—but that at the same time the Russian military is going without. The Russian military is sinking into obsolescence, while the Chinese military gets the most modern weapons in its inventory. To take just one example, Russia sold the Chinese Sukhoi aircraft that the Russian air force does not have.

This year the Kremlin announced plans to cut Russian ground forces in Siberia and the Far East by 20 percent. At the same time, each year it is selling $2 billion in high-tech weapons to China that Russia's own armed forces cannot afford.[39]

The Sino-Russian Border Dispute

For many years, China and Russia (or the Soviet Union) were at loggerheads with regard to their border. Based on treaties signed in the nineteenth century, the border generally followed the Amur and Ussuri Rivers in the RFE..Nevertheless, disputes arose because of Moscow's decision to occupy all of the islands after the Japanese occupied Manchuria in 1931. It was because of one of these small—and for practical purposes insignificant—islands that

China and the USSR clashed in March 1969. For a while it looked like the two countries might go to war over them. Fortunately, wiser heads prevailed, and both sides pulled back.

In 1991, following the collapse of the USSR, the two sides reached an agreement on the eastern section of the border. The agreement was ratified by the Russian Parliament in 1992, and two years later they agreed on the Western section of the border (with two minor exceptions). Later, both sides decided to demilitarize the border zone to a depth of 100 km, an action that hurt the Russian Army far more than the Chinese, because Beijing did not have many troops stationed near the border. Thus, for practical purposes, the issue of the 4,259-km-long border between China and Russia has been settled, even though some in the RFE strongly opposed the agreements because it meant "selling out Russian lands, and hurting territorial, economic and political rights of the local population."[40] Many in the RFE do not believe the territorial agreement is final. For them, conditioned by years of Soviet propaganda, it will only be a matter of time before the crush of Chinese population leads Beijing to seek more territory. Some 74 percent of the population in Primorskii krai expect China to annex all or part of the region "in the long run."[41]

Trade with China

With the collapse of the USSR, the unavailability of consumer goods, and the end of subsidies to the RFE, increased trade between China and the RFE became a necessity. By 1993, "trade between the Russian border regions and China's Heilongjiang province accounted for one-third of the overall Sino-Russian trade balance."[42] In some areas, Chinese and Russian border areas have developed very close economic relations. "In fact, the southern part of the Russian Far East and China's Donbei province formed an independent and complimentary organization."[43]

While there have been problems, e.g., the Chinese complain about mistreatment at the hands of Russian officials or the poor business practices of Russians, and the Russians express concern about shoddy Chinese goods and the tendency on the part of the Chinese to look down on the Russians, the fact is that both sides need each other. On the other hand, trade did not develop as well as both sides had hoped until after Putin took over. For example, while Yeltsin and Jiang Zemin pledged in 1996 to increase trade to $20 billion by 2000, "its volume has stagnated between $5 and $6 billion for the last three years."[44] One third of that is estimated to be border trade.[45] During 2000, however, it jumped 40 percent to $8 billion—not a lot when compared to U.S.-China trade, which amounts to $115 billion, but at least it is on the rise.[46] In late 2001, it was suggested that trade might rise to $10 billion.[47]

Russian-Asian Relations: The Putin Years

If there is one thing that has been completely consistent in Putin's policy toward the Far East, it has been his effort to use foreign ties as a way of solving the country's internal problems. This is not to suggest that enhanced Russian prestige is not important to him. Rather, whether it is the sale of weapons to China and India or the expansion of ties between the RFE and the countries in the region, almost everything he does in the foreign sphere has an internal political or economic purpose.

Indeed, in his public statements he has made it clear that Asia is of primary importance to the Kremlin. For example, speaking on January 26, 2001, he said,

> The Asian direction is gaining increasing significance. I will say it out-right: I believe it would be wrong to measure where we have more problems, in Europe or Asia. There must not be a Western or Eastern preference. The reality is that a power with such a geographical position as Russia has national interests everywhere. This line should be consistently pursued. We should get a firm foothold in all Asian affairs.[48]

In practice, this meant that Putin expected Russian diplomats to put on a full court press when it came to expanding Russian relations with Asian nations, especially those who were located on its borders and could play an important role in helping the RFE.

Japan

Moscow's relations with Japan remain stalled by the Kuril Islands dispute as in the past. The Kremlin would like nothing better than to resolve the issue, but neither side appears prepared to make the kind of compromise that would be required in order to solve it.

As a result largely of the Kuril Islands problem, relations between Japan and Russia have yet to be normalized. To take only one example, Japan's investment in the U.S. accounts for 42 percent of its total foreign investment; its investment in China is 7 percent of its foreign investment. But in Russia, Japanese investment is only 0.054 percent, despite the great need for investment in the RFE and Moscow's desire for foreign investment.[49] While the resolution of the Kuril Islands dispute would remove the major obstacle to

Japanese investment in the region, Japanese investors remain somewhat hesitant to jump into the RFE until a more attractive investment infrastructure has been created (for example, a legal system protecting investments).

As far as.Putin himself is concerned, he realizes how important resolution of this problem is, but he is also wise enough to know that he already has enough problems with opposition to his reform plans from conservatives and the military. As a result, he has tried to take the diplomatic high road without doing anything to end the dispute. During an interview in July 2000, he noted that "Japan and Russia are natural partners because we are neighbors. We need each other." Turning to the territorial dispute, he noted, "The main thing here, it seems to me, is to be patient, not to rush ahead but to work, as I have already said, on the basis of the legitimate interests of both sides." Or as he further noted, "I am absolutely convinced that if we work in this direction, if we develop cooperation in all areas, then the problems on the path to a peace treaty will cease to be so major and irritating."[50]

In many ways, the situation was as confusing as it was contradictory. On the one hand, the inhabitants of these four islands feel themselves to be Russians (the Japanese were expelled after the Russians arrived). On the other hand, the Russian Far East, and especially Primorskii krai, Khabarovsk krai, and Sakhalin oblast, depend on Japan for one third of their trade.[51] Furthermore, by resolving this dispute, Russia would be in a better position to deal with China, an emerging new superpower in Asia. In order to resolve this vexing problem, both sides will have to be ready to compromise.

The bottom line, however, is very simple. The Russians need the Japanese more than the Japanese need Russia. Furthermore, the Japanese have made it very clear that a major improvement in bilateral relations (including large-scale investment in the RFE) depends on resolution of the territorial dispute and the signing of a peace treaty. To quote a Japanese representative speaking in February 2001,

> If the leaders of Japan and Russia can resolve the difficulty lying between us and normalize our bilateral relationship on the basis of a peace treaty, the ties between the two people will become dramatically stronger. Such friendly Japan-Russia relations would greatly contribute to peace and stability in the Asia-Pacific region and, indeed, of the entire world.[52]

Putin continues to push for resolution of this dispute. During a press conference held when Japanese Prime Minister Mori visited Russia in 2001, he stated, "I consider it absolutely important, that we continue our dialog on the difficult aspects of our bilateral relations."[53] He went out of his way to mention the declaration of 1956, suggesting that he was prepared to make the same kind of deal that was suggested under Khrushchev—a partial return of two of the islands. Indeed, in December 2001, the Russian Foreign Ministry called on the Japanese to "compromise." The Russian Foreign Ministry reiterated Moscow's readiness to turn over to Japan the islands making up the lower Kuril chain, indicated as Shikotan and Habomai.[54] As late as February 2002, Russian Foreign Minister Igor Ivanov visited Tokyo, and the issue was discussed; but the Japanese again insisted on the return of all four islands.[55] The push for closer ties with Japan from the RFE combined with the weakened power of the Russian navy could help Putin push for a change on the Kuril Islands dispute. Much will depend, of course, on how strong his political position in Moscow is in the future.

China

China is the largest, and clearly the most important, country in Asia from a geographical standpoint. In addition, it remains the most important insofar as the Russian Far East is concerned, and is the major market for Russian weapons sales as noted above.

From the beginning, there were hints that while Putin recognized the importance of China, he did not want to tie the Kremlin too closely to China. For example, in spite of a Chinese request that he visit Beijing while he was still acting president, he put off a visit until two months after he had been formally inaugurated (Putin was elected in March 2000 and inaugurated in May 2000). He was given a red-carpet welcome when he arrived in Beijing on July 18, 2000. While the two sides signed a joint statement warning that Washington's effort to develop a theater missile system could upset stability in the region, the real focus—from the Russian standpoint—was on economic and military cooperation. Indeed, while the military-to-military relationship continued to grow as noted above, there was special focus on the economic integration of the RFE and China. To name only one example, a rail link between Primorskii krai and the Chinese province of Jilin was officially opened (after eight years of construction) just prior to Putin's arrival. While this rail line is only 40 kilometers long, it has the carrying capacity of up to 3 million tons of freight a year and was hailed by officials from Primorskii krai as a major breakthrough.[56]

In July 2001, Chinese President Jiang Zemin visited Moscow. The two sides signed a Treaty on Good Neighborly Friendship and Cooperation Between the Russian Federation and the People's Republic of China. On the one hand, the treaty sent a clear message to Washington that if it continued on its present path of unilateralism, the two sides would have no alternative but to move to a closer military relationship.[57] In spite of the polemical tone of the treaty, it was an empty diplomatic gesture. Nowhere did it state that one side would come to the other's aid in the event of military hostilities, nor did Moscow agree to assist Beijing militarily should the Taiwanese problem heat up again.[58]

From Putin's standpoint, the most important outcome of Jiang's visit was the signing of two trade deals. One was an agreement that Russia would sell China five Tupolev 204-120S passenger planes, and the second was a pipeline deal that had been talked about for several years. The proposed 1,500-mile pipeline would run from the Kovykta oil field in Siberia to China. The line would connect Angarsk, a Russian petrochemical center in southern Siberia, with the industrial regions in China's northern provinces. Construction of the pipeline is estimated at U.S. $1.7 billion. According to Russia's Energy Ministry, the line would carry 20 million tons of crude oil annually upon completion in 2005 and 30 million tons after 2010.[59]

Current plans call for a further expansion of economic ties between the two countries. For example, "experts predict that Russia will be able to export 25 billion to 30 billion cubic meters of natural gas to China annually, as well as 15 billion to 18 billion kilowatts of electricity from the newly completed hydro power stations in Siberia. . . . The two countries are also considering building a bridge over the Amur river to connect Heihe city in Heilongjiang province with Blagoveshchensk."[60] In addition, there are calls from Russian bankers for legislation that will permit Chinese financial institutions to operate in the Russian market.[61] In short, from all indications, in spite of the many problems of corruption and bureaucratic inefficiency, the message is that economic ties between Russia and China, and especially between Beijing and the bordering oblasts, will continue to intensify in the future.

Another area where the interests of the two countries coincide is opposition to Islamic radicalism in Central Asia. While it was originally set up in 1996 to resolve border disputes, the attention of the Shanghai Five (China, Russia, Tajikistan, Kazakstan, and Kyrgyzstan) soon shifted the fight against Islamic fundamentalism. In 2001, the group was expanded to include Uzbekistan. China, in particular, was concerned that its large and restive Muslim population in the western part of the country could become radicalized. To illustrate, the statement adopted in Dushanbe in July 2000 "pledges the . . . countries to jointly crack down on liberation movements, terrorism, and religious extremism in their borders."[62] And then at the June 14, 2001, summit in Shanghai, Presidents Putin and Jiang focused almost entirely on "issues of regionalism, particularly terrorism."[63] At this point, it remains unclear just how far cooperation will go between the six countries. While the Shanghai Six is a loose federation, it is worth noting that other countries such as Pakistan, Iran, India, Mongolia, and Turkmenistan have expressed an interest in joining. Furthermore, collaboration between the six will include joint military planning as well as technical assistance. If nothing else, this could be useful to both Beijing and Moscow as a vehicle for undermining U.S. influence in the region should Washington decide to keep troops in the region once the Afghanistan operation is over.

Conclusion

Putin has not made any secret of his desire to improve relations with the West, especially the U.S. For example, in March 2000, he surprised almost everyone by talking about the possibility of Russia's becoming a member of NATO, and by May 2002 it was clear that Russia would have the same rights as other NATO members in every area except military affairs.

Putin seized on the events of September 11 to line Moscow up behind the U.S. and the West. He even took the Russian high command on directly on September 24, 2002, when he blasted them for their obstructionism and failure to support the U.S. in Afghanistan. Russian behavior changed significantly in the aftermath of that meeting. Russian military cooperation from Moscow has been excellent since that time.

Pavel Felgenhauer is one of the best Russian analysts. Writing about Putin's policy toward the West, he noted, "The Kremlin understands, it seems, that only a close alliance with the West will afford Russia an opportunity to begin the country's technological and economic modernization, which has long been urgent."[64]

So where does this put the Far East when it comes to Russian foreign policy? Asia, and especially China and Japan, are critically important to Putin. China and India buy billions of dollars of weapons—money that Putin hopes to use to modernize the Russian defense establishment. At the same time, China also plays a very important role when it comes to supporting the economic health of the RFE. Without cross-border trade, the RFE would be in very serious trouble.

Putin appears frustrated with Russia's relationship with Japan. He knows that Japanese investment could make an important difference in the RFE, especially now that Moscow is making a serious effort to create the kind of legal and economic infrastructure necessary to sustain foreign investment.

In the meantime, Russia is pushing ahead with efforts at multilateral economic projects. Consider Sakhalin Island. Western oil giants like Royal Dutch/ Shell, Exxon-Mobil, and Chevron-Texaco "are preparing to invest $45 billion to help turn desolate Sakhalin into a massive oil-and-gas hub serving China, India, Korea and other parts of Asia for the next half century or longer."[65] While there are problems—Western officials must continue to deal with corrupt Russian officials—the fact is that the project is moving forward and will go a long way toward tying the RFE into the rest of Asia.

Internally, Putin is also working to improve the situation in the RFE. For example, during 2001, the Russian government decided to reform the defense industry sector. The reform has the most direct bearing on the Siberian region's interests. Approximately one-third of the Russian Federation's entire defense industry complex is concentrated here. The holding companies that are to be created will include around one-half of the military plants based in Siberia. The remaining defense enterprises . . . will not receive state orders and will either be converted to "civilian" output or will be disbanded.[66]

Most of the plants to be converted will focus on producing equipment for the Railways Ministry, the Ministry of Civil Defense, as well as for dealing with natural disasters, airports, and the mining industry.

Russian domestic and foreign policies intersect significantly in the Russian Far East, and this region will continue to play an important role in the Kremlin's Asian policy. Putin has made it clear that he expects to raise the Russian flag again in Asia. Toward this end, he will continue to push for a Russian presence in international and bilateral organizations in Asia, but he realizes that integrating the RFE into a stable, regional economic relationship is his first priority.

Notes

1. Alexander Sergounin, "Post-Communist Russia and Asia-Pacific: Changing Threat Perceptions," p. 4. The University of Melbourne (October 1998). Available at http://www.cerc.unimelb.edu.au/bulletin/oct98.htm.

2. The Far Eastern Federal District includes the Republic of Sakha, the Amur oblast, Primorskii krai, Khabarovsk krai, Magadan oblast, Kamchatka oblast, Sakhalin oblast, the Jewish Autonomous Oblast, the Koryak Autonomous Obkrug, and the Chukotka Autonomous Okrug.

3. Michael Wines, "Chinese Creating a New Vigor in Russian Far East," *The New York Times*, September 23, 2001, reprinted in Johnson's Russia List, September 23, 2001. Available at www.cdi.org/russia/johnson.

4. Felix Chang, "Unraveling of Russia's Far Eastern Power," *Orbis*, vol. 43, no. 2 (Spring 1999), pp. 260–261.

5. Wines, "Chinese Creating a New Vigor in Russian Far East."

6. Sherman Garnett, "The Russian Far East as a Factor in Russian-Chinese Relations," *SAIS Review*, vol. 16, no. 2 (Summer–Fall 1996), p. 7.

7. "Running for the Border," *Newsday*, August 31, 2001, pp. 1–4.

8. Ibid.

9. Ibid.

10. Sergei Sevastianov, "Russian Reforms: Implications for Security Policy and the Status of the Military in the Russian Far East," *The National Bureau of Asian Research*, vol. 11, no. 4 (1999), p. 3, essay 1. Available at http://nbr.org/publications/analysis/vol11no4/essay1.html.

11. Viacheslav Amirov, "Russia and East Asia: Problems and Prospects for Economic Cooperation," *Russian and Euro-Asian Bulletin*, vol. 10, no. 2 (March-April 2001), p. 8.

12. Ibid.

13. "Pulikovsii Loses More Than He Wins," *Russian Regional Report*, vol. 6, no. 29, August 22, 2001, p. 6. Available at RRR@topica.email-publisher.com.

14. Amirov, "Russia and East Asia," p. 9.

15. Sevastianov, "Russian Reforms," p. 5.

16. "Russian Arms, Military Exports to Top $4 Billion in 2001," Reuters, December 17, 2001, in *WNC Military Affairs*, December 18, 2001, p. 1. Available at wnc@apollo.fedworld.gov.

17. Ibid.

18. Chang, "Unraveling of Russia's Far Eastern Power," p. 274.

19. Sergounin, "Post-Communist Russia and Asia-Pacific: Available at http://www.cerc.unimelb.edu.au/bulletin/oct98.htm.

20. During an official visit to the Russian Far Eastern Military District during the summer of 2001, the author had an opportunity to visit Russian forces in Khabarovsk and Vladivostok. The situation was not good. For example, the current flagship of the Pacific Fleet, the *Admiral Vinogradov*, was the same flagship used by the Russians during their visit to San Diego in 1980. It was also clear that it was not being lived on, and we were told that it had been to sea only once during the last year and that was to India for a special celebration.

21. "Japan's Northern Territories," Foreign Ministry of Japan, no date. Available at www.mofa.go.jp/region/europe/russia/index.html.

22. Rajan Menon, "Japanese-Russia Relations and North-east Asian Security," *Survival*, vol. 38, no. 2 (Summer 1996), p. 67.

23. Dmitri Trenin, "The Far Eastern Backyard," in *The End of Eurasia*. Available at http://pubs.carnegie.ru/english/books/2001/03dt/chp5.asp.

24. "Relations with the United States and Neighboring Countries," Ministry of Foreign Affairs of the Government of Japan, February 2001. Available at www.mofa.go.jp/region/europe/russia/index.html.

25. See Charles E. Ziegler, "Russo-Japanese Relations: A New Start for the Twenty-First Post-Communism Century?" *Problems of Post-Communism*, vol. 46, no. 3 (May/June 1999), p. 18.

26. "Japan-Russia Forge Ties at Sea," *Vladivostok News*, October 5, 1998, p. 1; "Russia, Japan Hook Up Hot Line," *Vladivostok News*, August 20, 1999, pp. 1–2.

27. See "Programma uglublenia sotrudnichestva v torgovo-ekonomicheskoi oblasti mezhdu Rossiiskoi Federatsii i Iaponii," September 5, 2000, from the website of the President of the Russian Federation. Available at www.government.gov.ru/english/statVP_engl_1.html.

28. As cited in ibid.

29. This staged approach is taken from Li Jingjie, "Pillars of the Sino-Russian Partnership," *Orbis*, vol. 44, no. 4 (Fall 2000), pp. 527–530.

30. This is discussed in Elizabeth Wishnek, "Prospects for the Sino-Russian Partnership: Views from Moscow and the Russian Far East," *The Journal of East Asian Affairs*, vol. 12, no. 2 (Summer-Fall 1998), pp. 421–422.

31. Evgenii Bazhanov, "Russian Perspectives on China's Foreign Policy and Military Development," unpublished manuscript, no date, p. 81.

32. See, for example, John Wilson Lewis and Xue Litai, "China's Search for a Modern Air Force," *International Security*, vol. 24, no. 1 (Summer 1999), pp. 84–85.

33. Sharif Shuja, "Moscow's Asia Policy," *Contemporary Review*, vol. 272, no. 1587 (April 1998), p. 4.

34. Bazhanov, "Russian Perspectives on China's Foreign Policy and Military Development," p. 81.

35. Stephen Blank, "Which Way for Sino-Russian Relations?" *Orbis*, vol. 43, no. 3 (Summer 1999), pp. 345–360.

36. Elizabeth Wishnick, "One Asia Policy or Two? Moscow and the Russian Far East Debate: Russia's Engagement in Asia," *IREX Scene Setter*, September 5, 2001, p. 1.

37. Michael Barron, "China's Strategic Modernization: The Russian Connection," *Parameters*, vol. 31, no. 4 (Winter 2001–2002), p. 73.

38. "Russia to Build Two More Destroyers for China," *The Jamestown Monitor*, January 10, 2002, p. 4. Available at brdcst@jamestown.org.

39. "Russian Xenophobia Threats to Sour Accord with China," July 14, 2001. Available at smb.com.au.world.

40. As cited in Bazhanov, "Russian Perspectives on China's Foreign Policy and Military Development," p. 83.

41. *The Financial Times*, August 27, 2001, p. 2.

42. Wishnek, "Prospects for the Sino-Russian Partnership," p. 430.

43. Sergounin, "Post-Communist Russia and Asia-Pacific," p. 8.

44. Wishnick, "One Asia Policy or Two?" p. 1.

45. "Running for the Border," *Newsday*, August 31, 2001, p. 1.

46. "Russia and China Boost Trade," *The Jamestown Monitor*, September 24, 2001, p. 1. Available at brdcst@jamestown.org.

47. "Chinese Creating a New Vigor in Russian Far East," the *New York Times*, September 23, 2001, p. 23.

48. Full text of President Putin's speech in the Russian Foreign Ministry on January 26, 2001, pp. 15–19, from Johnson's Russia List, January 29, 2001. Available at www.cdi.org/russia/johnson.

49. "Russian Territorial Governor Says 4 Islands Should be Returned," *The Yomiuri Shimbun*, May 17, 2001, p. 1.

50. "Key Quotes from Putin Interview," Reuters, July 12, 2000, in Johnson's Russia List, July 12, 2000. Available at www.cdi.org/russia/johnson.

51. Ibid.

52. Speech by Dr. Tatsuo Arima, the Representative Government of Japan at the 37th Munich Security Council Meeting: "Bilateral Relations with Russia from Japan's Perspective," February 4, 2001, Japanese Foreign Ministry website. Available at www.mofa.go.jp/region/europe/russia/index.html.

53. "Vystuplenie Prezidenta Rossiiskoi Federatsii V.V. Putina i otvety na voprosy na sovmestnoi press-konferentsii po itogam vsrechi s Premier-ministrom Iaponii Eo. Mori," March 25, 2001, Official Website of the President of the Russian Federation. Available at www.government.gov.ru/english/statVP_engl_1.html.

54. "Russia Urges Japan to Compromise on Disputed Islands," *ITAR-TASS*, November 29, 2001, in *WNC Military Affairs,* December 3, 2001. Available at wnc@apollo.fedworld.gov.

55. "Russian Foreign Minister in Tokyo," *The Jamestown Monitor*, February 3, 2000, p. 3. Available at brdcst@jamestown.org.

56. See Dmitry Bulgakov, "Sino-Soviet Ties That Bind," *The Russia Journal*, July 22–28, 2000, p. 1.

57. The Treaty was ratified by the Duma on December 26, 2001. See "Russian

Duma Ratifies Cooperation Treaty with China," Moscow Interfax December 26, 2001, in *WNC Military Affairs*, December 27, 2001, p. 1. Available at wnc@apollo. fedworld.gov.

58. By the end of 2001, the situation did not appear to be improving at the highest level. For example, the visit by Chinese Vice President Hu Jintao to Moscow on September 11 was handled in a very low-key manner. "Chinese Leader-in-Waiting Visits Moscow," *The Jamestown Monitor*, October 31, 2001, p. 2. Available at brdcst@jamestown.org.

59. "Russia and China Boost Trade," *The Jamestown Monitor*, September 24, 2001, p. 3. Available at brdcst@jamestown.org.

60. Ariel Cohen, "The Russia-China Friendship and Cooperation Treaty: A Strategic Shift in Eurasia?" *Heritage Foundation Backgrounder*, no. 1459 (July 28, 2001), p. 5.

61. "Far East Seeks Banking Ties with China," *Russian Regional Report*, vol. 7, no. 12, March 27, 2002. Available at RRR@topica.email-publisher.com.

62. "Shanghai Five: An Attempt to Counter U.S. Influence in Asia?" *Newsweek Korea*, May 2001, pp. 1–2.

63. "Putin, Jiang Open Shanghai Five Summit," *Associated Press*, June 14, 2001.

64. "Washington Is Putting Russia in Its Place," *Moskovskie Novosti*, December 18, 2001, in *WNC Military Affairs*, December 26, 2001, p. 1. Available at wnc@apollo. fedworld.gov.

65. Benjamin Fulford, "Energy's Eastern Front," *Forbes Magazine*, December 10, 2001, reprinted in Johnson's Russia List, December 24, 2001, p. 9. Available at www.cdi.org/russia/johnson.

66. "Siberian FD, Siberian Accord Debate Reform of Region's Defense Industry," *Rossiiskaia Gazeta*, December 21, 2001, in *WNC Military Affairs*, December 26, 2001, p. 1. Available at wnc@apollo.fedworld.gov.

For Further Reading

Elizabeth Wishnick, "Prospects for the Sino-Russian Partnership: Views from Moscow and the Russian Far East," *The Journal of East-Asian Affairs*, vol. 12, no. 2 (Summer–Fall 1998), pp. 418–451.

Alexander Sergounin, "Post-Communist Russia and Asia-Pacific: Changing Threat Perceptions," The University of Melbourne (October 1998). Available at http://www.cerc.unimelb.edu.au/bulletin/oct98.htm.

Felix Chang, "Unraveling of Russia's Far Eastern Power," *Orbis*, vol. 43, no. 2 (Spring 1999), pp. 257–284.

Charles E. Ziegler, "Russo-Japanese Relations: A New Start for the Twenty-first Century?" *Problems of Post-Communism*, vol. 46, no. 3 (May/June 1999), pp. 15–25.

Stephen Blank, "Which Way for Sino-Russian Relations?" *Orbis*, vol. 43, no. 3 (Summer 1999), pp. 345–360.

Sergei Sevastianov, "Russian Reforms: Implications for the Security Policy and the Status of the Military in the Russian Far East," *The National Bureau of Asian Research*, vol. 11, no. 4 (1999), essay 1. Available at http://www.nbr.org/publications/analysis/vol11no4/essay1.html.

Andrei Piontkovsky, "Does Russia Want to Keep Its Far East?" *The Russia Journal*, August 23–30, 2001, pp. 5–8.

3

The Challenge of Relations with Former Republics

Mikhail A. Alexseev

To understand post–Soviet Russia's security challenges in the former Soviet republics, imagine that history at the end of the twentieth century took a different turn and that capitalism and democracy underwent a deep crisis, leading to disintegration of the United States, Canada, and Great Britain. Imagine that California, Texas, Florida, and eleven other American states, as well as Quebec, British Columbia, Northern Ireland, and Scotland, seceded from these countries and became internationally recognized independent states almost overnight, with seats in the United Nations. Imagine that some of these newly independent states retained their own nuclear arsenals and that in some the new governments adopted hostile policies toward their respective "old countries" to retaliate for alleged colonialism and oppression. Moreover, imagine that these secessions were not just "civilized divorces" with larger states splintering into smaller ones, but that territorial disintegration was part of the wholesale political and economic transformation of the U.S., Canada, and Great Britain. In fact, imagine that the Washingtonians, the Anglo-Saxon Canadians, and the English—paradoxically as it would seem—supported secessions of their peripheries because they themselves wanted to secede, except that their secessions would not be territorial, but political, as they rushed to escape the tyranny of checks and balances, the stock market, and the International Monetary Fund (IMF). At the street level, imagine that having smashed the Washington Monument, the Toronto Tower, and the Nelson Column, the pro-communist and pro-central-planning crowds also demanded the dissolution of the costly and inhumane capitalist democratic "empires" and the liberation of fellow oppressed citizens in California and Scotland. NATO, the European Union, and NAFTA automatically disintegrated. Had all this indeed happened, the United States, Canada, and Great Britain would have to be replaced by new states, with new political systems, new economies, and new identities.

But this would be only part of the story. Imagine also that as these up-
heavals cast capitalism and democracy onto the scrap heap of history, the
Soviet economy flourished, Soviet society got increasingly more vibrant and
lively, and the Soviet military and economic bloc—the Warsaw Treaty Organization and the Council for Mutual Economic Assistance (COMECON)
—expanded to the whole of continental Europe, Greenland, Mexico, Central America, and the Caribbean. And imagine that the Soviet Union demonstrated its overwhelming military superiority and global reach precisely
when capitalism and democracy in the West were under mortal challenge
from domestic protesters with a massive and successful military intervention reversing a hypothetical takeover of Venezuela by, say, Colombia—in
the same fashion as the United States reversed the takeover of Kuwait by
Iraq in the 1991 Gulf War. As capitalism crumbled, crowds in the heart of
Moscow around Red Square celebrated the Soviet victory in the Cold War.
Furthermore, suppose that shortly after this, the Soviet Union—as the only
remaining superpower—decided to pursue a strategy of communist enlargement with the westward expansion of the Warsaw Treaty Organization and
the COMECON with an eye on including California, Texas, and Scotland
and without ruling out membership of other "newly independent states" in
the future. Finally, imagine that against the background of these tectonic
shifts, civil wars erupted in the former states and provinces of the United
States, Canada, and Britain as diverse ethnic populations grew insecure
and more hostile toward one another in the near anarchical environment
following the collapse of capitalism. Unstable, fear-producing environments
throughout the post-capitalist world engendered ethnic cleansing, refugee
flows, poverty, aggressive religious fundamentalism, banditry, warlordism,
and terrorism.

Reversed Anarchy: Russia, the "Near Abroad," and
International Relations Theory

To understand Russia's post-Soviet identity crisis, what debates the new security challenges engendered, and how specific responses to these challenges
emerged, I suggest assessing Russia's security situation in the context of a
theoretical innovation that I define as "reversed anarchy" or "structural anarchy reversed."[1]

Defined as the absence of central government authority, anarchy is considered to be the organizing principle of relations among sovereign states
in the tradition of political realism. Mainstream international relations theories implicitly concur that there is more anarchy outside than inside states.
The idea of "reversed anarchy" is disarmingly simple—one faces more

anarchy on the inside than on the outside of a state. Reversed anarchy would obtain when (a) a government that fears state failure and related challenges (such as the ones resulting from post-Soviet multilevel identity crisis) interacts with (b) a group of powerful and increasingly interdependent states and their institutions, especially when (c) the latter are located (or operate) within close proximity to the internally challenged state. "Fear" in this context implies that perceptions matter decisively. To paraphrase Alexander Wendt, reversed anarchy is what decision-makers make of it.[2] Reversed anarchy is hardly something that characterizes all politics at the global level. At the same time, uneven evolution of domestic and international order and disorder suggests that reversed anarchy perceptions are most likely within internally challenged states coming into contact with increasingly interdependent international actors.[3]

Russia and the former Soviet republics in the 1990s represent the case where all three elements of reversed anarchy have been evident. Internally weakened as a result of the wholesale breakdown of Soviet political and economic institutions, Russia faced an existential challenge to territorial integrity with the deadly wars breaking out in Moldova, Georgia, Azerbaijan, Tajikistan, and, ultimately, on Russia's own territory, in Chechnya. Claiming greater powers from Moscow in a sweeping "parade of sovereignties" of the early 1990s, the constituent regions and republics of the federation challenged Russia's state identity. Precisely at that time, Europe was consolidating and expanding its international institutions to Russia's borders,[4] while European and Asian economies grew more integrated with the economy of the United States—the state that acquired a *de facto* monopoly on global security in the post–Cold War world. Moscow had good reasons to see Russia as hopelessly peripheral to the grand game of economic globalization.[5]

Reversed anarchy enables one to systematically address the relationships among the domestic, "intermestic," and international dimensions of the identity crisis engendering insecurity in post-communist, post–Cold War Russia and to understand the role of the "near abroad" in Russia's foreign and security policy in the 1990s and beyond. The focal point of security in this situation is the *linkages* between internal and external sources of threat.[6] Addressing such linked threats for post–Soviet Russia meant combining strategies of state consolidation at the domestic level; various forms of reintegration and/ or increasing policy leverage at the "intermestic" ("near abroad") level; and positioning of Russia as a Eurasian "great power" at the international level. In this complex calculus, successful state consolidation, i.e., prevention of Russia's complete or partial disintegration, is pivotal both for increasing its policy leverage in the "near abroad" and for making at least a credible appearance that Russia could be a viable player internationally. Consolidation of Russia's influence in the former Soviet republics, in turn, helps reduce domestic challenges to sovereignty and enhances Russia's international stature. And having greater stature and leverage in the capitals of the most powerful nation-states and in international organizations helps to consolidate both the Russian state and the Russian position as the leading actor in the former Soviet area. Russian policy makers and analysts also understood the price of failing to achieve these goals which, under a worst-case scenario, would represent the ultimate threat to Moscow—that Russia would become weak, isolated, marginalized, and poor.

Thus, under reversed anarchy the Russian government had to maintain multiple-policy equilibria and balance among various policies geared at simultaneously increasing domestic consolidation, "near abroad" leverage, and international stature. As Aleksei Arbatov, deputy chairman of the defense committee of Russia's lower house of parliament, the State Duma, wrote in 1997, Moscow's major dilemma in this regard was "how to find the right balance between treating the other former Soviet republics as independent, and cultivating some kind of 'special relationship' with them."[7] The Kremlin policies toward the "near abroad" have not consistently conformed with either the "revanchist" agenda (consistent with realism) of the Communists and Zhirinovsky Nationalists, or the reformist agenda (consistent with neoliberalism).[8] Focusing on linkages binding internal and external challenges to security under "reversed anarchy" explains the emergence of extreme policy positions in the CIS in the early 1990s, their coalescence into compromises in the late 1990s, and the wholesale re-evaluation of security challenges in the "near abroad" after the terrorist attacks on the United States on September 11, 2001.

Trends and Developments from Yeltsin to Putin

The evolution of Russia's policies toward the Commonwealth of Independent States (CIS) in the first post-Soviet decade is characterized by three major paradigm shifts keyed by changes in Moscow's strategies to deal with the country's identity crisis under reversed anarchy. The shifts occurred as each of these strategies encountered costly obstacles and generated negative unintended consequences. The predominant focus was first on international (1991–1993), then on intermestic (1994–1998), and finally on domestic challenges (1999–2001) to Russia's security and identity. The search for an optimal three-level balance continues to define Russia's "near abroad" policy at the time of this writing, with an emphasis again shifting toward international interactions in the post–September 11, 2001, security environment.

Phase One: "Joining the Civilized World" to Resolve the "Near Abroad" Challenges

From 1991 to the end of 1993, Moscow emphasized international-level solutions to the post-Soviet identity crisis within Russia and the "near abroad." The Kremlin sought to reinstate Russia's status as the world's "great power" not through aggrandizing Russia's territorial reach in Eurasia—as had been the pattern since Russia emerged from the Tatar-Mongol rule in the early 1500s—but through cooperation and integration with rich industrial democracies. This strategy was consistent with the "new thinking" philosophy of the last Soviet leader, Mikhail Gorbachev, which abandoned Soviet expansionist ambitions and sought "a return to civilization." The first Yeltsin government that assumed power in early 1992 argued that the costs of maintaining Soviet-style economic and political influence in the former Soviet republics would be detrimental to democratic, free-market reforms and integration with the West. As Leon Aron suggests, the neoliberals in the first Yeltsin government viewed integration and cooperation with the West as important both to "harmonize Russian interests and those of the 'civilized world' (as capitalist democracies often were referred to in the Russian media), but also to blunt and, eventually, extirpate ancient ethnic animosities on the territory of the former Soviet Union."[9]

In this sense, "harmonization" between Russia and the West presented itself as the best method to avoid Balkan-style armed conflicts and minimize social turmoil within and around Russia, and to exercise benign, yet critical, political and economic influence in the "near abroad." Moscow at the time came to share the paramount Western security concern linking external and internal sources of threat—the replay of the Yugoslavia-style violent civil war across the former Soviet Union with nuclear weapons. Since these issues were inextricably linked to the legacy of Soviet ethnofederalism and Soviet global military ambition, the new government in the Kremlin had a clear incentive to dissociate itself from these legacies. The Russian leaders assumed that common perception of post–Cold War global threats with Washington—such as nuclear proliferation, internal wars, terrorism, and religious fundamentalism—would induce the West to underwrite Russia's political and economic transformation and ensure benign international participation in preventing and/or ending violent conflicts in the former Soviet Union.

Pursuing this strategy, the Yeltsin government insisted on transferring control over nuclear weapons to the CIS—as an institutional vehicle for eventually centralizing nuclear weapons control under Russia. Cooperation between Russia, the United States, and international institutions such as the Atomic Energy Agency culminated by the mid-1990s in the transfer of Soviet nuclear weapons to Russia's control from Ukraine, Belarus, and Kazakhstan. In 1993, Russia capitalized on its new strategic partnership relationship with the United States in engineering a U.S.-Russia-Ukraine deal under which Ukraine agreed to transfer its nuclear weapons to Russia in exchange for Russia's writing off most of Ukraine's nearly $1 billion debt for natural gas and the U.S. agreeing to purchase from Russia reprocessed nuclear materials extracted from Ukrainian weapons. In other developments signaling Moscow's abandonment of imperial designs, Russia's foreign minister, Andrei Kozyrev, outlined policies seeking to (1) withdraw Russian troops from the former Soviet republics as fast as possible; (2) support handing over control of the former Soviet military forces in the newly independent states to their central governments; (3) support international recognition of the central governments of the CIS states and withhold support from separatist opposition, such as the Russian movement in the Transnistria area of Moldova; and (4) invite outside powers to mediate and settle conflicts in the former Soviet republics, including the use of the UN and even NATO forces in ethnic conflict zones.[10] The speaker of the Russian parliament, Ruslan Khasbulatov, in February 1992 warned against "dragging Russia into inter-ethnic conflicts [in the former USSR]," not even in a mediating role. "Russia," said Khasbulatov, "should abandon the practice of regarding all its initiatives as history-making." Instead, the Russian parliament speaker called for "the triumph of pragmatic approaches to Russia's foreign political priorities," meaning that Russia should give priority to "the promotion of contacts with those countries and can help Russia in real terms build up its economic might."[11]

In the economic arena, Russian liberals saw the CIS as a tool to preserve mutually beneficial Soviet-era economic ties among the former republics, but without paying Soviet-style subsidies to them. For this reason, independent Russia's first prime minister, Yegor Gaidar, in January 1992 argued for the abolition of the "ruble zone" in the CIS on the grounds that a common currency under political independence would create the incentive in the former republics to unilaterally increase local salaries, thus engendering a run on the currency and runaway inflation. As a matter of general principle, according to the Russian parliament deputy and economist, Vasilii Seliunin, "Anyone thinking of joining the Commonwealth in order to get subsidies and handouts must be told right away that they are knocking on the wrong door, that there are no pies to be divided here, that there is nothing to get from the common barrel, that here there are peoples of sovereign states exchanging the fruits of their labors to mutual advantage."[12] By default, this position signaled Moscow's abandonment of "neocolonial" ambitions in the "near abroad" (i.e., maintaining its capacity to dictate policy through economic pressure). In addition, while shedding the "costs of the empire" in the "near abroad," Russia actively sought billions of dollars in loans and assistance funds from international financial institutions, rich industrialized states, multinational corporations, private financiers, foundations, and other nongovernmental organizations. The internationalist emphasis, however, would soon have to be scaled down.

Phase Two: Enhancing "Intermestic" Security

Moscow's predominantly liberal strategy combining international cooperation in the "far abroad" with isolationism in most of the "near abroad" soon encountered powerful external constraints and domestic challenges. While hopes that the West would undertake the equivalent of the Marshall Plan to integrate Russia into the world economy alongside rich industrialized democracies remained illusory, Moscow had to deal with massive economic downturn and political violence in the "near abroad," and the Kremlin could not conceivably wait for Western assistance to help resolve the burning problems close to home. Further, Moscow lost any illusion that it could successfully resolve violent conflicts in the CIS in cooperation with the major Western powers. In other words, Moscow realized it lacked the leverage to convert its cooperation with increasingly cohesive strong states in the "far abroad" into assuaging anarchy in the former Soviet borderlands.

Russia's internationalist liberal policy legitimated sporadic and often disorderly withdrawal of former Soviet interior ministry and army troops from the Chechen republic in the North Caucasus and from Azerbaijan, including Nagorno-Karabakh (a predominantly Armenian-populated enclave in Azerbaijan fighting for independence with Armenian support). According to Pikayev, "Russia's political self-restraint had sent the wrong signals to some capitals [in the former Soviet republics, such as Moldova], which interpreted them as a green light for using force against secessionists."[13] In the absence of a central authority able to deter or suppress an aggressive contestant, conflicts in Georgia (Abkhazia and South Ossetia), Moldova (Transnistria), Azerbaijan (Nagorno-Karabakh), and Tajikistan escalated to mass violence, resulting in approximately 139,500 deaths and 2,414,000 refugees by 1994.[14] To return to my "reversed analogy" with North America, imagine that conflicts erupted in newly independent Texas, Florida, and Arizona claiming equally massive casualties and think of the impact these conflicts would have on the public and politicians in the remaining "core" states. Moreover, with 25 million ethnic Russians and an estimated 42 million Russian-speakers "beached" in the former republics as the result of the dissolution of the USSR, the Kremlin could vividly visualize scenarios when Russia would be flooded by millions of people seeking refuge in their ethnic and linguistic homeland from violence in the "near abroad." Given the long history and strong economic and social ties of Russians with the CIS states, these conflicts quickly became a hot domestic issue, enabling the meteoric rise to political stardom of nationalists and populists ranging from the more moderate Alexander Lebed to the radical Vladimir Zhirinovsky.

At the same time, the majority of Russians (rather simplistically, but understandably) came to associate Russia's economic decline—inflation, layoffs, unpaid wages—with the Western involvement in the "shock therapy" style of economic reforms, such as privatization. Over 60 percent of Russians polled by American political scientist Jerry Hough in late 1995 said the West was "pursuing the goal of weakening Russia with its economic advice."[15] Economic hardship and political anarchy (*bespredel*) also undermined support for democratic reforms. In a University of Strathclyde survey of 2,426 Russians in January 1996, only 9 percent chose "democracy" while 77 percent preferred "order" when asked which of the two was "more important for Russia."[16]

These pressures necessitated major revisions in Moscow's CIS strategy. According to a review of this strategy by the Moscow State Institute of International Relations (MGIMO, a school of diplomacy under Russia's foreign ministry), "the question about relations with the Soviet successor states became the target of heated debates, during which powerful political groups criticized the government for underestimating the role of the 'near abroad' and insisted on having a special relationship with the CIS states. They argued that these states comprised the sphere of Russia's special interests and Russia's influence exclusive of outside powers."[17] The strategy changes were reflected in the revised version of the Russian foreign policy concept signed into law by President Yeltsin in April 1993. The document outlined two principal foreign and security policy goals—"strengthening positive relations with the CIS states in order to prevent processes of disintegration on the territory of the former Soviet Union" and "retaining Russia's position as a great power." The concept was called, *inter alia*, for "maximum integration" of the former Soviet republics, ensuring that Russia remained the only nuclear power in the former Soviet area, retention of Russia's central role in the defense of the CIS external borders, and prevention of armed conflicts in the "near abroad." Russia saw itself as cooperating with the United Nations and a rather amorphous Conference on Security and Cooperation in Europe (CSCE, currently OSCE), but not with NATO, in preventing or resolving these conflicts. Domestic opposition to NATO's enlargement later translated into fierce aversion of NATO's involvement in any of the post-Soviet hot spots.

The perception of increasing domestic anarchy against the background of increasing political and economic cohesiveness of rich industrialized states around Russia—to Russia's exclusion—cued the Kremlin's threat assessments and policy in the "near abroad" and "far abroad." In *The White Book of Russia's Special Services*—perhaps the most comprehensive articulation of the Russian intelligence community's view on national security in the mid-1990s[18]—the growing Western power and capacity to expand its institutions was directly posited as a security threat to Russia in the CIS: "Despite changes in the last several years, *the principles of power balance and power interaction remain* the core of contemporary international relations [original emphasis]. . . . The United States now includes regions of the former USSR—the Baltics, the Transcaucasus, the states of Central Asia—in its sphere of vital national interests."[19] The ultimate threat was envisioned, in that with that support behind them, "practically all neighbors of Russia are likely to make territorial claims if Russia's domestic crisis worsens."[20] The MGIMO assessment likewise stated that at the time "strengthening of the influence of any of the third parties in this [post-Soviet] area (in particular, Turkey's influence in Central Asia and the Transcaucasus, Romania's influence in Moldova, Poland's influence in Belarus and Ukraine, and the U.S. and Western influence in the Caspian region, Ukraine, and Central Asia) was interpreted as something that automatically weakened Russia's influence and ran contrary to Russia's vital interests."[21]

However, given multiple-issue linkages at the domestic, intermestic, and international levels, Russia's strategy that crystallized in the September 1995 document *Strategic Course of Russia Toward the CIS States* followed a middle course between *Realpolitik* and liberal institutionalism. After all, the Yeltsin government continued to need Western credits, technical assistance, and deferral on repayment of over $100 billion of Soviet-era debt. Rather than pushing for restoration of Soviet or Russian imperial institutions throughout the former Soviet states, the Kremlin visualized the strengthening of the CIS as a new center of regional interdependence and, in that sense, a global player. In other words, the Kremlin chose competitive regional interdependence as a response to reversed anarchy challenges. Hence, the *Strategic Course* envisioned rapid CIS-wide institutionalization. Regarding the economy, this strategy called for the establishment of the Customs Union, the Payments Clearing Union, the Currency Union, model legislation (not dissimilar from the Council of Europe and European Union model documents), a common capital market, transnational financial-industrial groups, and, ultimately, a CIS free-trade area (CIS Economic Union). Regarding security, the *Course* emphasized the conversion of the Tashkent collective security treaty concluded on May 15, 1992, into "a system of collective security" leading prospectively to "a defense union." Concurrently, Moscow established the goal of "ensuring that CIS states do not participate in unions or blocs working against any of the member states." The Kremlin wanted to institutionalize joint border service commands to guard the CIS external borders and to intensify cooperation among CIS intelligence services.[22]

Moscow took it for granted, however, that institutionalization of the CIS would be a vehicle for raising Russia's own international stature and for stealing the thunder from the vociferous nationalist and communist opposition at home. CIS integration policies were stepped up after the arrival of Yevgenii Primakov, former academic and foreign intelligence chief, as Russia's foreign minister in January 1996. Primakov stressed that strengthening Russia's position in the CIS would a top priority in Russia's foreign policy. His first official visits were to Turkmenistan and Ukraine. Primakov made a powerful symbolic gesture toward strengthening ties with Ukraine by flying there in a small, uncomfortable, noisy Antonov-24 turboprop designed in Kiev and used extensively for air travel within the CIS. He later appointed one of Russia's most distinguished diplomats, Yurii Dubinin, as Ambassador to Ukraine.[23] Sections in Russia's foreign ministry dealing with the Soviet successor states were upgraded to the status of ministerial departments, with Primakov presiding over regular meetings of all CIS ambassadors.[24]

During the fiercely contested presidential campaign in 1996 in which he started as a distant underdog with about 5 percent approval ratings in December 1995, Yeltsin spared no effort, portraying himself as an "integrator" of the former Soviet republics. At the same time, Yeltsin needed to set himself apart from his Communist and nationalist rivals, who pursued integrationist policies through the Russian parliament, the State Duma. Hence, Yeltsin immediately and vigorously denounced the Duma resolution, passed in March 1996, that invalidated the 1991 Belaia Vezha accords on the dissolution of the USSR. Yeltsin explained his opposition to this resolution, however, by stating that it could undermine the *de facto* reintegration process and justify the claim of former Soviet satellites in East and Central Europe for full NATO membership.[25] In other words, by denouncing the Duma resolution, Yeltsin aimed at scoring points with the voters for being a better, more realistic integrationist of the former Soviet republics. Having blasted the Duma, Yeltsin produced, within less than a month, the Integration Agreement between Russia, Belarus, Kazakhstan, and Kyrgyzstan, which envisaged creating "supranational political institutions" to regulate the wide range of economic and military issues.[26]

Phase Three: "State Consolidation" at the International-Domestic Frontier

This strategy of promoting Russia's position as a global and regional power through trans-CIS political, economic, and security institutions also ran into insurmountable obstacles. With Russia lacking both the resources to attract or pressure former Soviet states and a reputation for benign multilateralism, the former Soviet republics proved reluctant to engage more than symbolically in the proposed institutions and common projects. Taking stock of Russia's CIS policy by 1999, the MGIMO review concluded: "Other than counting a few tactical successes, none of the integrationist goals set in the principal CIS policy documents was achieved by the late 1990s."

From an economic standpoint, CIS states have been a sizable, but peripheral, trading partner for Russia. The share of the CIS states in Russia's foreign trade increased from 17.8 percent in 1992 to 25.9 percent in 1993, but after that it steadily declined to 21.6 percent in 1998. In contrast, 50 percent of Russia's foreign trade in this period was with Western Europe. Russia ceased to be the principal loan provider to the CIS states—by 1997 CIS states outside Russia were receiving $25 billion in loans from the "far abroad" exceeding several times the amount of loans granted by Russia. The customs, payment, currency, and free-trade unions did not materialize, with the exception of the customs union between Russia and Belarus.[27] In the security arena, only six CIS states (Russia, Armenia, Belarus, Kazakhstan, Kyrgyzstan, and Tajikistan) remained part of the CIS collective security agreement in 1999, while Azerbaijan, Georgia, and Uzbekistan opted out of the treaty. The CIS defense union failed to materialize, and there were disputes over funding of CIS peacekeepers and ex-Soviet military installations, including missile attack early-warning systems in Kazakhstan, Tajikistan, Azerbaijan, Ukraine, and Belarus, the Baikonur space center in Kazakhstan, and border troops in Georgia, Kyrgyzstan, and Turkmenistan. At the same time, all CIS states joined NATO's Partnerships for Peace program and signed bilateral military cooperation agreements outside the CIS.[28]

Meanwhile, while most violent armed conflicts within the CIS subsided by 1994, that same year Moscow had to deal with the consequences of ethnopolitical separatism and its own failure at conflict prevention within the Russian Federation itself. Following the "parade of sovereignties," when most of Russia's 89 constituent regions and republics claimed political and economic authority at Moscow's expense, Russia got bogged down in a brutal, persistent, and intractable military conflict in Chechnya starting in December 1994. The first war (1994–1996), which ended after Moscow withdrew its troops under overwhelming domestic pressure that nearly cost Yeltsin re-election to the second term in the summer of 1996, claimed an estimated 30,000 civilian casualties and displaced 600,000 persons.[29] The Russian military was humiliated and had to withdraw, while the newly elected Chechen government declared the republic "subject to international law." With Chechnya potentially becoming independent, Moscow became concerned that the secession of Chechnya would galvanize Islamic-based separatist movements elsewhere in the North Caucasus, especially in Dagestan.

Thus, after 1997 Moscow saw increasing anarchy on the inside in contrast to increasing interdependence on the outside, including the growing cooperation between states in the "near abroad" and "far abroad," with the Baltic states seeking NATO and EU membership and Georgia and Azerbaijan suggesting they could join NATO as well. Center-periphery relations in Russia —especially concerning Chechnya and the North Caucasus—then fed into what may be termed a "negative feedback loop." Perceptions of growing external interdependence (from expansion of Western institutions to penetration of Russia and the CIS by Islamic fundamentalist networks) accentuated perceptions of internal vulnerabilities.

The 2000 Foreign Policy Concept directly associates domestic challenges to Russian sovereignty with increasing interdependence of major Western states: "Integration processes, in particular in the Euro-Atlantic region, often tend to be selective and limited. Attempts to downplay the state sovereignty as the fundamental element of international relations pose the threat of unconstrained interference in internal affairs." The concept further emphasized that "strengthening of Russia's statehood"—meaning stronger central authority—would be decisive for Russia to "gain a dignified position in the world."[30]

In this way, perceptions of international security became inseparable from perceptions of the state of center-periphery relations within Russia. Putin's Center for Strategic Research argued that by the year 2000 the increased capacity of Russian regions to influence the federal government "posed real threats to sovereignty and integrity of the Russian Federation." Moreover, "in their scope and embedded risks, these threats have become more significant than the recently [January 1999] adopted concept of national security stated they were."[31] The report called for resolving center-periphery and national security issues on the basis of "a comprehensive approach" (*kompleksnyi podhod*). The "comprehensive approach" logic explains why Moscow constructed linkages between external interdependence and internal security, overestimating the negative implications of Western policies.

For example, the Gulf War came to be interpreted not as an instance of beneficial international cooperation to prevent interstate wars and territorial claims by force, but as an instance of the United States and its former Cold War allies acquiring license to intervene wherever they have economic interests.[32] If the Gulf War could happen over Kuwait, it could—in the view of the head of the North Caucasus military district expressed privately in 1998—plausibly happen over Chechnya and Dagestan.[33] Similarly, if NATO mustered resolve to intervene in Kosovo and to help Islamic pro-independence fighters, it could muster resolve to intervene in the North Caucasus and boost Islamic pro-independence fighters there. Or at least, such actions were seen as able to generate nonmilitary external support to domestic secessionists and put pressure on the central government to grant Chechnya independence. Moreover, in a security dilemma, perceived threats escalate rapidly and nonmilitary involvement of outside powers then plausibly appears to serve as a prelude to military intervention. Hence, Putin's uncompromising opposition to Western mediation in internal disputes over sovereignty between Moscow and the regions: "We were offered intermediaries to resolve the conflict in Chechnya. We do not need any intermediaries there. That would be the first step toward internationalizing the conflict. First, intermediaries, then somebody else, then observers, then military observers, and then things will be out of our control."[34] Similarly, the Russian foreign policy concept states: "Attempts to introduce into the international lexicon concepts such as 'humanitarian intervention' and 'limited sovereignty' in order to justify the use of force without UN authorization are inadmissible."[35]

The threat of Russia's being "squeezed out" of the North and South Caucasus (and, hence, out of the "great game" for the Caspian oil resources) if Chechnya achieved independence contributed to the perception of the clear and present threat of Russia's economic disintegration in the late 1990s. Moscow at the time was alarmed about the capacity of regional governments to impede interregional commerce through quasi customs offices and embargoes, to issue regional quasi currencies (usually in the form of IOUs), to buy off representatives of federal agencies such as the border and custom services, and to collude with financial-industrial groups on tax evasion and price fixing.[36] This threat was particularly acute after the 1998 financial and currency crisis and the devaluation of the ruble.

Overall, these perceptions meant that state consolidation of Russia itself took precedence over creating supranational institutions in the CIS under Russia's leadership. Thus, according to the MGIMO assessment, the theme of CIS integration in Russia's elite political discourses begins to ebb starting in 1998. Starting in 1999 Russia ceased to insist that CIS institutions must be granted authority above that of newly independent states. In the revised version of the *Concept of Russia's National Security* signed into law by President Vladimir Putin on January 10, 2000, cooperation with CIS states is no longer listed among the top priority goals. Of the twelve "main tasks" of Russian foreign policy, the *Concept* identifies only one (number five) as "the development of relations with the member states of the Commonwealth of Independent States according to the principles of the international law and promotion of those integration processes within the CIS that meet Russia's interests."[37]

A New Phase: CIS, Terrorism, and the Post–September 11 World

In the late 1990s, Chechnya—where Russia launched another war in 1999 that continues at the time of this writing, having claimed more than 3,500 Russian military lives—became the focal point of linkages between external and internal threats to Russia's security. Chechnya was seen not only as a rebellious province, but as the gate through which anarchy in the "near abroad" (intermestic level) was destined to spill into Russia's domestic politics and trigger Russia's disintegration. Commenting on these perceptions in November 2000, Dmitri Trenin of the Moscow Carnegie Center wrote:

> To paraphrase a major work of political writing, a specter is haunting the minds of Russian decision-makers, the specter of Islamic extremism. They see a plot aimed at rolling back secularism along a wide arc (or front) stretching from Kosovo across the Caucasus and Central Asia all the way to the Philippines. This phrase, first used by General Vladislav Sherstyuk, a deputy secretary of the Russian Security Council, was subsequently adopted by President Putin and is now being repeated by lower officials.[38]

A year before the terrorist attacks on New York and Washington, the Russian Federal Security Service produced a report alleging that Afghanistan-based Osama bin Laden and Al Qaeda masterminded a plot to carve independent quasi states out of Muslim-populated areas of Russia (in Chechnya, Dagestan, Tatarstan, Bashkortostan, and Astrakhan) and out of the newly independent Central Asian states and merge them into a "Great Islamic Caliphate." Citing instability in Tajikistan and Uzbekistan, the "Islamic nuclear bomb" obtained by unstable Pakistan, the revival of Islam within the Russian Federation, and external funding for Islamic extremists in Russia (allegedly up to $1.3 billion from January to September 2000), the report made a point, as Trenin put it, that "Russia will not only suffer at the edges: the territorial unity of the federation will be irrevocably compromised."[39]

The acuteness of this threat perception is also illustrated in the writings of General Gennadii Troshev, one of Russian commanders in the Chechen military campaigns. According to Troshev, Chechen rebel fighters under the command of the Jordanian Amir Khattab received training in Pakistan; Pakistan, Turkey, and Saudi Arabia financed and encouraged the fundamentalist "Islamic Society of Dagestan" ("Jamaat") to fight for a joint Islamic state with Chechnya and for secession from Russia; Saudi Arabia financed Chechen separatists to thwart pipeline projects transferring Caspian oil to world markets through the Caucasus; the Taliban leader Mullah Omar blessed the attack of Chechen fighters under Shamil Basaev and Amir Khattab on Dagestan;

and Osama bin Laden not only transferred $30 million to Basaev and Khattab, but personally visited terrorist training camps near the Chechen village of Serzhen-Yurt.[40] Concerns raised in Western Europe and the United States about the brutality of Russian military campaigns in Chechnya were interpreted then as a signal that the rich industrialized nations were *de facto* supporting instability within Russia.

However erroneous, these perceptions explain the momentous turnaround in Russian policy toward the West and NATO after the September 11 attacks in the United States. Whereas prior to September 11, Russian elites viewed the United States and the West as potential beneficiaries of domestic and "near abroad" turmoil, the change of U.S. foreign policy toward the war on "terrorism with global reach" meant that Russia again had a chance to benefit decisively from international cooperation to minimize security concerns in the CIS and domestically. Moreover, entry into the antiterrorist alliance with the United States signaled greater prospects to Moscow for getting a substantive part in decision-making within major international institutions, such as NATO, and for entry into the World Trade Organization. This became clear immediately after President George W. Bush's embrace of Russia as an ally in the war on terrorism in response to Vladimir Putin's being among the first leaders who expressed condolences after the New York and Washington attacks. (Thinly veiled in these condolences was the point that Russia had recognized the international terrorism threat and sought alliance with the U.S. but had been ignored until tragedy struck at home). With Russia as a potential major U.S. global ally, the incentive to perceive the U.S. and Western institutions such as NATO as posing a threat to domestic stability and security in Russia diminished. According to the chairman of the defense and security committee at Russia's upper house of parliament, Viktor Ozerov, after September 11, 2001, "the problem of Chechnya as it appeared to the West before is no longer on the agenda."[41] Significantly, President Putin agreed to the United States's dispatching 200 military advisers to Georgia to help the local military deal with terrorist groups circulating between Chechnya and northeastern Georgia's Pankisi Gorge.

With the nature of perceived threats and opportunities transformed, Moscow could not only offer Washington the comfort of canceling missile tests at the time of heightened security alerts, but also provided rights to use Russian air space, endorsed the setting up of U.S. military bases in the former Central Asian republics, and supplied weapons to the anti-Taliban Northern Alliance in the fall of 2001. Predictably, combating terrorism was the first item on the agenda of the CIS summit meeting on February 28, 2002, in Almaty, Kazakhstan. As a *Nezavisimaia Gazeta* observer commented, "Over the past four months, the geopolitical environment has changed so substantially that President George Bush could be an appropriate guest at the summit meeting of CIS leaders. While the CIS was being formed, none of its participants could have imagined in their worst nightmares that NATO aviation bases would be located in Kyrgyzstan, Uzbekistan, and Tajikistan, feeling almost at home here; nor that some 'Western guests' would arrive in Georgia."[42]

Conclusion: After Reversed Anarchy

The perspective of reversed anarchy provides a succinct explanation both of the change in direction of Russia's policies in the "near abroad" and—more counterintuitively—of the nature and surprising endurance of the Commonwealth of Independent States as an institution. Counter to the *realpolitik* predictions, Russia did not abandon the CIS to pursue "Russia first" policies nor did it convert the CIS into an instrument for restoring a Soviet or imperial type of institutional domination over the former republics. Counter to the neoliberal predictions, Russia did not convert the CIS into an instrument of cooperative integration of the former Soviet republics and itself into expanding Western institutions, especially in Europe. Understandably, in Russia the CIS—and the Russian policies in the "near abroad" that it embodies—has been a disappointment both to the Communists and Nationalists, on the one hand, and to the liberal internationalists on the other. In the West, the CIS, unsurprisingly, has been regarded first as a short-lived, and later as an ineffective, organization. But while the CIS has failed the *realpolitik* and neoliberalist expectations, it has served and may potentially continue to serve a larger purpose to mitigate the challenges arising from fear-producing environments under reversed anarchy. The CIS principal security role, from this standpoint, was not to generate the EU-style Defense or Border Union, and not to become Moscow's tool for dictating policies in Kiev or Tbilisi, imperial-style, but to help avoid the replication of deadly disputes among the former Soviet republics in a context not dissimilar to the one observed in the former Yugoslavia. The CIS became effective not because it achieved any policy outcomes, but because it persisted and could therefore serve as a bargaining forum where conflicting intentions could be signaled to multiple parties and misperceptions leading to mistrust and possibly violence softened. The CIS gave breathing space to post-Soviet states to work on their respective identity and security challenges. It has served as a forum that Moscow-wary republics could treat as a vehicle to greater independence and Moscow could treat as a tool for decreasing threats originating at the international, intermestic, and domestic levels simultaneously. As a *realpolitik* or a neoliberal institution the CIS was ineffective and inconsequential, but as an institution arising to address the challenges of reversed anarchy it has been pivotal to assuage and most likely prevent the repetition of the Yugoslavia war scenario across the nuclear-armed former USSR.

As for likely future developments, Putin's perception that his policies to rein in Russia's provincial leaders since 2000 have been successful and that the threat of state disintegration has been avoided, coupled with the change in the post-September 11 international environment, suggest that Russia is likely to see the former Soviet republics less as a source of threat to Russia's integrity and international stature and more as an arena of cooperation with major Western states and international institutions. This strategy is particularly attractive because it offers Russia a chance to resolve its post-Soviet identity crisis and emerge in the global arena as a cohesive state and a vital player in international institutions expanding into still "anarchic" parts of the world, such as the Caucasus and Central Asia. Not surprisingly, by March 2002, President Putin was reported to have been working on major foreign policy changes. These included Russian troop withdrawal from Georgia, new policies toward Moldova, a "nonmilitaristic approach" to Chechnya, and making the most out of the opportunity to speedily join the World Trade Organization. According to Liliia Shevtsova of the Carnegie Endowment for International Peace, with these changes "Russia also is trying to redefine its identity to consolidate its society and a new role on the international scene."[43] If this combination of domestic and external trends holds and Russia succeeds in capitalizing on them, the reversed anarchy challenges will diminish, and complex interdependence is likely to be a better guide to understanding post–September 11 CIS policies.

Notes

1. Mikhail A. Alexseev, "Regionalism in Russia's Foreign Policy in the 1990s: A Case of 'Reversed Anarchy,'" *The Donald W. Treadgold Papers*, no. 37 (Seattle, WA: Henry M. Jackson School of International Studies, 2001).

2. Alexander Wendt, "Anarchy Is What States Make of It: The Social Construction of Power Politics," *International Organization*, vol. 46, no. 2 (Spring 1992), pp. 391–425.

3. A similar phenomenon has been suggested for other parts of the world. "While in modernity the inside of a state was supposed to be orderly, thanks to the workings of the state as a Hobbesian 'Leviathan,' the outside remained anarchic. For many states in the Third World, the opposite seems closer to reality—with fairly orderly relations to the outside in the form of diplomatic representations, but total anarchy within." See Bjorn Moller, "The Role of Military Power in the Third Millennium," in Charles W. Kegley, Jr., and Eugene R. Wittkopf, eds., *The Global Agenda: Issues and Perspectives* (Boston: McGraw-Hill, 2001), p. 46. This idea, however, is not developed further in the article. In the work to which Moller refers, R. B. J. Walker, *Inside/Outside: International Relations as Political Theory* (New

York: Cambridge University Press, 1993), the author discusses this situation as the "inside/outside" problem in international relations that goes down to the definition of individual vs. state identity in relation to the international system. Neither Moller nor Walker, however, systematically analyzes international security implications of reversed anarchy as defined in this chapter, or of the inside/outside problem more generally.

4. This has been particularly evident in the adoption of a common currency by the European Union, NATO's going "out of theater" in the former Yugoslavia, and in EU and NATO ongoing enlargement. In fact, even without eastward expansion, continuation of NATO after the end of the Cold War—in contrast with the disbandment of the Warsaw Pact—would be a powerful indicator to Moscow of institutional cohesiveness on the outside of the former Soviet area at the time of Soviet disintegration and rising challenges to the Kremlin on the inside.

5. For example, U.S. direct investment into Western Europe was $48 billion in 1994 and $72 billion in 1999; in Asia and the Pacific it exceeded $13 billion in 1994 and $29 billion in 1999. However, in Russia and all Soviet successor states, U.S. direct investment was less than $1 billion a year during the same period. See U.S. Department of Commerce, Bureau of Economic Analysis (2000), *U.S. Direct Investment Abroad: Capital Flows, 1994; U.S. Direct Investment Abroad: Capital Outflows, 1999*, available at (http://ww.bea.doc.gov).

6. The implication here is that chiefs of government of State A would not necessarily be threatened more if State B amassed greater or more destructive military capabilities, but if State B is associated with international institutions that may be used to intervene in domestic conflicts of State A or in conflicts on State A's periphery. Reversed anarchy would then explain why the enlargement of NATO into Eastern Europe would be perceived as a greater security threat by Moscow than a nuclear weapons buildup by India, Pakistan, and even Iran and Iraq. And that would be despite the obvious calculus that an accidental or hostile nuclear launch targeting Russia would be more likely to come from new and politically unstable nuclear powers to Russia's south than from NATO states.

7. Alexei Arbatov, "Russian Security Interests and Dilemmas: An Agenda for the Future," in Alexei Arbatov, et al., *Managing Conflict in the Former Soviet Union: Russian and American Perspectives* (Cambridge, MA: The MIT Press, 1997), p. 415.

8. Some authors have labeled this group as "internationalists," as in Leon Aron, "The Foreign Policy Doctrine of Postcommunist Russia and Its Domestic Context," in Michael Mandelbaum, ed., *The New Russian Foreign Policy* (New York: Council on Foreign Relations, 1998), pp. 23–64. Others have labeled this group as "isolationists," as in Alexander Pikayev, "The Russian Domestic Debate on Policy Toward the 'Near Abroad,'" in Lena Jonson and Clive Archer, eds., *Peacekeeping and the Role of Russia in Eurasia* (Boulder, CO: Westview Press, 1996), pp. 51–66. The term "neoliberals" suggests that both labels are consistent with this group's ideological outlook in that they viewed Russia's top priority as abandoning Soviet expansionism (hence, "isolationists") and simultaneously in joining the "club" of advanced free-market democracies to regain international stature (hence, "internationalists").

9. Aron, "The Foreign Policy Doctrine," p. 23.

10. Andrei Kozyrev, "Rossiia. God minuvshii i god nastupivshii," *Diplomaticheskii vestnik*, nos. 1–2 (1993), pp. 3–5.

11. BBC Summary of World Broadcasts, "Yeltsin and Kozyrev Outline Russia's

Foreign Policy in the New World Order," February 28, 1992, on Lexis-Nexis. Available at http://web.lexis-nexis.com/universe.

12. Vasilii Seliunin (trans., Michael Vale), "Not the Sword, but Peace!" *Russian Social Science Review*, vol. 34, no. 1 (January–February 1993), pp. 44–52, Internet (EBSCO) version. Available at http://search.epnet.com/.

13. Pikayev, "The Russian Domestic Debate," p. 57.

14. Ted Robert Gurr, "People Versus States: Ethnopolitical Conflict and the Changing World System," *International Studies Quarterly*, vol. 38, no. 3 (September 1994), pp. 369–375; *State of World Conflict Report, 1995–1996* (Atlanta: The Carter Center, 1996), p. 50.

15. Jamestown Foundation, *Prism*, vol. 2, no. 2, part 2 (June 2, 1996). Available at (www.jamestown.org/).

16. Michael Kramer, "The People Choose," *Time* (May 27, 1996), p. 56.

17. A. V. Zagorskii, *Rossiia i SNG*, in A. V. Torkunov et al., eds., *Vneshniia politika Rossiiskoi Federatsii 1992–1999* (Moscow: ROSSPEN, 2000), p. 97.

18. *Belaia kniga Rossiiskikh spetssluzhb* (Moscow: Obozrevatel', 1995). This collaborative project featured 22 editorial advisors and 133 contributors, including the public relations chiefs of both the Foreign Intelligence Service (formerly the First Chief Directorate of the KGB) and the domestic Federal Security Service (FSB), and post-Soviet Russian intelligence chiefs, Viktor Barannikov and Sergei Stepashin. The latter served as Russia's prime minister and was succeeded by Vladimir Putin in August 1999.

19. Ibid., pp. 13–14.

20. Ibid., p. 15.

21. Zagorskii, *Rossiia i SNG*, p. 102.

22. Ibid., p. 101.

23. *RFE/RL Newsline*, part I (May 27, 1996). Archival material available at www.rferl.org/newsline/search/.

24. *Vremia*, Russian Public Television News, July 29, 1996, 1:00 PST.

25. *RFE/RL Newsline*, part I (March 19, 1996). Archival material available at www.rferl.org/newsline/search/.

26. *RFE/RL Newsline*, part I (April 1, 1996); ibid. (March 29, 1996). Archival material available at www.rferl.org/newsline/search/. Following the election victory, Yeltsin capitalized on the commonality of his position on the "near abroad" with the Communists by appointing a high-ranking communist, Aman Tuleyev, to the newly established position of minister for cooperation with CIS Member States. Tuleyev accepted, saying he found no discrepancies between his party and the Yeltsin administration's views on CIS integration. Tuleyev's first practical steps called for: merging the CIS-member energy systems into a single network; increasing Russia's access to the oil and gas resources of Kazakhstan, Azerbaijan, and Turkmenistan; and putting CIS states with debts to Russia into *de facto* receivership through a massive Russian acquisition of their industrial equity. See *RFE/RL Newsline*, part I (September 10, 1996). Archival material available at www.rferl.org/newsline/search/.

27. Zagorskii, *Rossiia i SNG*, pp. 103–4.

28. Ibid., pp. 105–6.

29. Estimates by the U.S. Holocaust Museum.

30. Ministry of Foreign Affairs of the Russian Federation, *Kontseptsiia vneshnei politiki Rossiiskoi Federatsii,* Moscow, June 28, 2000. Available at www.scrf.gov.ru/Documents/Decree/2000/07–10.html.

31. B.A. Bagrandzhiia, "Vliianie sub"ektov RF na reshenie voprosov federal'nogo znacheniia." Moscow: Center for Strategic Research, 2000. Available at www.csr.ru/conferences/bagr.html.

32. *Belaia kniga Rossiyskikh spetssluzhb*, p. 18.

33. Reported at the U.S. Department of State conference, "Prospects for the North Caucasus," Meridian House, Washington, DC, April 7, 1998.

34. Vladimir V. Putin, *Ot pervogo litsa: razgovory s Vladimirom Putinym* (Moscow: Vagrius, 2000), p. 158.

35. *Kontseptsiia vneshnei politiki*. Available at www.scrf.gov.ru/Documents/Decree/2000/07–10.html.

36. Daniel S. Treisman, "Russia's Taxing Problem," *Foreign Policy*, no. 112 (Fall 1998), pp. 55–66.

37. Quoted in Zagorskii, *Rossiia i SNG*, p. 110.

38. Dmitri Trenin, "Central Asia's Stability and Russia's Security," Program on New Approaches to Russian Security, Policy Memo no. 168 (Washington, DC: Center for Strategic and International Studies, 2000), p. 3.

39. Trenin, "Central Asia's Stability," pp. 3–4.

40. Gennadii Troshev, *Moia voina: dnevnik okopnogo generala* (Moscow: Vagrius, 2001), pp. 170–71, 174–75.

41. RosBusiness Consulting Database, "NATO Official Thinks Disagreements Between Russia and NATO Become Unimportant," November 23, 2001, on Lexis-Nexis. Available at web.lexis-nexis.com/universe.

42. *Nezavisimaia Gazeta*, March 1, 2002, p. 3, on Lexis-Nexis. Available at web.lexis-nexis.com/universe.

43. United Press International, "Russian Foreign Policy Revolution?" in Johnson's Russia List, no. 6132, March 13, 2002. Available at www.cdi.org/russia/johnson.

For Further Reading

Lena Jonson and Clive Archer, eds. *Peacekeeping and the Role of Russia in Eurasia* (Boulder, CO: Westview Press, 1996).

Alexei Arbatov, et al., *Managing Conflict in the Former Soviet Union: Russian and American Perspectives* (Cambridge, MA: The MIT Press, 1997).

Zbigniew Brzezinski and Paige Bryan Sullivan, *Russia and the Commonwealth of Independent States: Documents, Data, and Analysis* (Armonk, N.Y.: M.E. Sharpe, 1997).

Mark A. Smith, *Russian Foreign Policy 2000: The Near Abroad* (Camberley, Surrey: Conflict Studies Research Center, Royal Military Academy Sandhurst, 2000).

A. V. Torkunov, et al., eds., *Vneshniia politika Rossiiskoi Federatsii 1992–1999* [The foreign policy of the Russian Federation, 1992–1999] (Moscow: ROSSPEN, 2000).

4

The Challenge of Russia's North-South Integration

Gregory Gleason

At the time the USSR disintegrated, Russian policy makers were under the assumption that Russia would continue to play a leading role throughout the countries of southern Eurasia even after the end of the USSR. Russia had long played a leading role in southern Eurasia. Russia's claim to the lands of the North Caucasus dates back to Ivan the Terrible's capture of Astrakhan in 1556. Russia expanded into Kazakhstan and western Siberia in the 1850s, captured key Chechen warlords in 1865, and pushed through Central Asia into Afghanistan in the 1870s. After the collapse of the Russian monarchy in 1917, Russian Bolsheviks quickly announced their intention to retain Tsarist Russia's hegemony over the Caucasus and Central Asian regions by bringing them into the socialist fold.

The seven decades of Soviet control of southern Eurasia witnessed the transformation of many aspects of life, from commerce, science, and industry to culture itself. While efforts to "Sovietize" the countries of the region were by no means completely successful, the Russian stamp of the Soviet period was unmistakable.[1] By the late 1980s, virtually all the important commercial ties of the region had been oriented northward for over a century. The entire professional and managerial class of the countries had been Russified, with many of them having been trained in the universities and technical institutes of Moscow and the other cities of the Eurasian north. The Russian language was the *lingua franca* of the realm, commanded by anyone who had aspirations to succeed. Although the Soviet Union claimed to be an internationalist state, in fact, the southern Eurasian region was largely under Russian control.

The Russian presence in the Caucasus and Central Asia during the Soviet period had many critics among the indigenous peoples. At the same time, Russia's interactions with the countries of the Caucasus and Central Asia had critics within the Russian political elite. Russian political economists had long been arguing that the Central Asian countries and the Caucasus were economic liabilities, supported at the expense of the north Eurasian regions.[2] Before the breakup of the Soviet Union, one of Mikhail Gorbachev's most popular political themes had been the pro-Western—and thus anti-Eastern—appeal for the restoration of a "common European home." By the time the USSR began disintegrating in 1991, many of Russia's most important politicians greeted the general Russian withdrawal from the regions of the Caucasus and Central Asia with enthusiasm and relief.

But Moscow policy makers who welcomed shedding the burden of directly administering the southern Soviet republics also assumed that Russia would retain its historical position of hegemony in the region. The hastily crafted Alma-Ata Declaration and the other agreements that established the Commonwealth of Independent States (CIS) in December 1991 reflected the assumption of the Russian representatives that maintaining a level playing field in commerce and policy would leave relatively advanced Russia in a privileged position.[3] Russian policy makers thus naturally expected that Russia would pursue a unified policy with respect to the Eurasian "near abroad" that was located in what Russians typically referred to as "the East"—the Caucasus, Kazakhstan, and Central Asia. Many political leaders felt that Russian political control had contributed little to Russia's security. Russian diplomats reasoned that Russia's security interests along its southern frontiers could be adequately ensured through diplomacy and treaty obligations.[4] As the disintegration of the USSR proceeded and millions of ethnic Russians were stranded as second-class citizens outside their homelands, the Russian government did little to relieve their distress, anticipating that it was only a matter of time before Russian hegemony returned to south Eurasia.[5]

Subsequent events proved these assumptions to be unwarranted. As the post-communist transition proceeded, Russia's capacity to exert decisive influence over the regions of Central Eurasia receded. Independent-minded states such as Georgia and Uzbekistan challenged Russia's claim to a dominant voice in the security and economic arrangements in the region. Russia's own internal political disputes, shrinking government revenues, the Russian financial markets crisis in 1998, weak world market commodity prices, and the rising European and U.S. influence throughout the Eurasian region hampered Russia's ability to maintain Moscow's former sphere of influence in the Caucasus and Central Asia. Russia lost the capacity to influence infrastructure development, as international organizations and large multilateral lending institutions moved into the region, establishing new mechanisms and new priorities for economic development.

Thus, as the first decade of post-Soviet independence proceeded, Russian foreign policy makers found their hands tied and their choices crimped. On occasion, Russia attempted to retain control over south Eurasian regional politics and oil development through intimidation, coercion, and surreptitious support for coups and armed conflicts in Azerbaijan and Georgia, by the invasion of Chechnya in 1994, by securing basing rights in Armenia, Georgia, and Turkmenistan, by blocking the transport of Kazakhstan oil and gas exports, and by retaliating against independent-minded Uzbekistan through establishing a Russian military outpost in neighboring Tajikistan. These heavy-handed Russian attempts to regain control were usually unproductive and often counterproductive, inclining their south Eurasian partners to seek stratagems of self-help and greater independence.

The Russian assumption of a grand policy toward Central Asia and the Caucasus gradually devolved into a situation in which Russia was conducting numerous parallel but not always complementary bilateral foreign policies. Sometimes this put Russia in a position of being able to divide the countries to its benefit, but more often it simply meant that none of the countries succeeded in even the most basic forms of infrastructure and commercial cooperation. The list of failed cooperation attempts speaks volumes: the CIS itself, the CIS Collective Security Treaty, the Central Asian Union, the Black Sea Forum, the Belarus-Russian Union, the Minsk Group, the Caucasus Four, the Caspian Five, the Central Asian Cooperation Organization, the Shanghai Five, the Shanghai Cooperation Organization, and the Eurasian Economic Community.[6] Eventually, dwindling intra-regional trade, the failure of international policy harmonization, and a growing concern with threats of insurgency and lawlessness gradually persuaded Moscow to acknowledge that its approach to Central Asia was fragmented, ad hoc, and unsuccessful.[7] By the late 1990s, a consensus had emerged within the Russian foreign policy community that the time had come for a new strategy toward the southern Eurasian states.

This chapter surveys the security and economic challenges of Russia's involvement with its southern neighbors, arguing that after 2000 Russia adopted a new posture toward the Eurasian states that emphasized economic integration and international policy coordination as against heavy-handed coercion and intimidation. The chapter argues that the goal of Moscow's new strategy was to allow Moscow to gain and exercise control over the management of key resources, particularly gas, oil, and transportation infrastructure. Three factors were particularly important in forging the consensus for the new policy. The first was the consequences of the Afghanistan war, the Chechen war, and the rise of insurgency movements in southern Russia, the Caucasus, and Central Asia. The second factor was the competition for commercial advantage, particularly in gas and oil markets. This approach led to new enthusiasm among the Moscow leadership for "soft hegemony" in southern Eurasia, carried out through the instrumentalities of economic integration and interstate policy conformance. Finally, the third key factor in Russia's evolving approach to south Eurasia is the dramatic shift in the regional balance of power that took place following the American-led anti-Taliban war in Afghanistan.

Russian Reforms and the Transformation in Eurasia

In foreign affairs, Russia sought to ensure its long-term economic and security interests through reliance upon the CIS, the amorphous "cooperation organization" that Russian diplomats assumed would encourage cooperation through mutually reinforcing bonds of self-interest. In domestic affairs, Russia adopted a development strategy closely modeled on European political experience and heavily emphasizing a Eurocentric orientation. Western-style democratic political institutions were established.[8] Russia undertook macroeconomic reforms, liberalizing prices, adopting a convertible currency, privatizing services and industry, and reconfiguring property relations along lines that would lead the country toward a European-style economy and polity. European legislative and regulatory structures, including a civil code mimicking that of the Netherlands, were grafted onto Soviet-era legislation. Administrative and bureaucratic structures were quickly adopted to bring the country up to European international standards.

It is true that a communist economy and polity was transformed in a shorter period of time than most reformers and communists thought possible.[9] But it is also true that the outcome of the foreign and domestic reforms was less than satisfactory. Steep declines in economic output, financial turmoil, and political infighting led to economic contraction to the point that by 2000 Russia's economy was only about 80 percent the size of that of the Netherlands, a dismal indicator that masks what should have been a windfall positive effect of the threefold increase in world oil prices between 1999 and 2001. The shortcomings of Russia's strategy of simultaneous political and economic liberalization invited unfavorable comparison with China's comparatively successful policy of economic modernization *sans* political liberalization. The deteriorating standard of living for a sizable proportion of the Russian population, corruption scandals at the highest levels of government, the rising profile of Russian oligarchs, the threats of international crime syndicates and narco-business, and finally the rise of ethnic separatism and terrorism shaped a political environment of popular dissatisfaction and distrust in government.

Russia's internal disarray and its diminishing international influence contributed to the incapacity of the Russian government to act decisively in the
last period of the Yeltsin years. But as new leadership began to emerge from
within the palace intrigues of the Russian bureaucracy, a new political ethic
also emerged. Russian leaders began to stress institutional reform and foreign relations that put greater emphasis on Russian historical experience, the
historical role of the state in Russia's economy, Russian traditions of person-
alism and strong-willed leadership, Russia's Asiatic roots and traditions as
opposed to European traditions, and the importance of Russia's foreign relations with Asian powers. By the time Yeltsin appointed Putin as acting president on January 1, 2000, a consensus had already formed in the Ministry of
Foreign Affairs that some new approach toward south Eurasia was necessary. Putin's public statements in late 1999 had already noted that Russia
was, after all, a "Eurasian power." These statements had set the stage for a
reexamination of Russia's strategy toward Eurasia. The "Russian National
Security Strategy" of January 2000 and the "Russian Foreign Policy Strategy" adopted in July 2000 formalized the reassessment of Russia's foreign
policy interests.[10]

Russia's new foreign policy received its formal initiation in an extensive
round of diplomatic exchanges that took place in spring and summer 2000,
as the Putin government undertook a systematic effort to articulate its new
"Asia policy." Judging from the government's rhetorical statements and the
content of the numerous joint communiqués signed during the May-July 2000
round of diplomatic meetings in Moscow, Tashkent, Dushanbe, Beijing,
Pyongyang, and Okinawa during this period, the Putin government was seeking to reestablish its posture in Asia. At the same time, a complementary
policy was being developed with respect to southern Eurasia, that is, toward
the Caucasus countries of Armenia, Azerbaijan, and Georgia and the Central
Asian countries of Kazakhstan and Kyrgyzstan, Tajikistan, Turkmenistan,
and Uzbekistan. Once openly described as "younger brothers," the newly
independent states of the Caucasus and Central Asia were regarded in the
new policy formulations as "partners." The new partnership was designed to
take into account the fact that the countries of the region were very different
and that to avoid a fragmented strategy, Russia would have to shape a policy
that would take into account these differences.

Russia, the Caucasus, Central Asia, and the Shadow of
Afghanistan

The assumption of leading Russian policy makers in the early period following the Soviet collapse that Russia would continue to be the dominant power

in the Eurasian region had profound implications for foreign policy calculations among the Eurasian states. The assumption led to a number of rarely discussed but firmly held expectations about Russia's future policy in the region. First, it was assumed that Russia would eventually return. Russia remained a nuclear power and the inheritor of much of the strategic capability of the USSR. Russia's dominant position fueled a fear that at some point Moscow could act to restore aspects of the Soviet system.

The speed of Uzbekistan's de-Sovietization was to some extent predicated on an attempt to make recidivism impossible by eliminating Russian influence as quickly and thoroughly as possible even at the expense of the disruption of commerce, science, and industry. Second, it was assumed that Russia could and would derail long-term arrangements for the transportation of Caspian-region oil and gas that did not directly benefit Russia. Accordingly, the other Caspian countries sought to find alternative routes that would eventually diminish Russia's bargaining position by diminishing Russia's transport monopoly. Third, it was assumed that Russian policy was motivated, if not to encourage regional animosities over Afghanistan, at least to seek to benefit from the disorder to the south of Uzbekistan's and Tajikistan's borders. Fourth, it was assumed that Russia would never be satisfied with the delimitation of the Russia/Kazakhstan border. Disenfranchisement of the Russian minority community in Kazakhstan would be an irritant to Russia that, under some circumstances, could call forth Russian nationalism in a way that would threaten Kazakhstan's territorial integrity. Accordingly, Uzbekistan could afford to be nationalistic, but Kazakhstan was constrained to foster "Eurasian-ness" rather than nationalism.

During the early days of the disintegration of the USSR, the foreign policy calculations of all the post-Soviet states of the region—Armenia, Azerbaijan, Georgia, Kazakhstan, Kyrgyzstan, Uzbekistan, Tajikistan, and Turkmenistan— were predicated upon a principle of national independence *outside* the Union of Soviet Socialist Republics but *within* the sphere of influence of Russia. No matter how deep the resentments and antagonisms of the Soviet period and no matter how strong the urge for full restoration of national autonomy, the states of the Central Eurasian region continued to base foreign policy decisions on the assumption that Russia remained at center stage in post-communist reorganization. Russia reciprocated by asserting the principle that special relationships were to be maintained with the former Soviet countries. But each of the Eurasian states had features and circumstances that meant the special relationships all would differ from one another.

What Russian policy makers had not calculated on was the continuing influence of the outcome of the war in Afghanistan. The Soviet Union had first invaded Afghanistan in December 1979 in an effort to pacify its southern border, prevent the contagion of Islamic extremism from spreading into Soviet territory, and carry out a socialist-style "progressive" reform.[11] The Soviet occupation provoked fierce opposition, galvanizing the various regional and ethnic factions into unified opposition against the Kabul-based government. Afghanistan's seven main factions united in the *jihad* against the occupation. Invoking the long-standing tradition of guerilla warfare against foreign invaders, the Mujahedeen fighters sought and received assistance from the outside world. Western powers, led by the U.S., were anxious to see the USSR's aims for territorial expansion contained by an indigenous opposition. The U.S. Congress provided security aid to the Mujahedeen. Most of the assistance was funneled through Pakistan, particularly to the opposition encampments filled with Afghan refugees in Pakistan's northeastern province.

Afghanistan proved to be an intractable conflict for the USSR and was one of the important factors in the dissension that swept Soviet society in the 1980s, leading to the disintegration of the country. After years of futile conflict, the Soviet Union resolved in 1986 to withdraw. The April 1988 Geneva accord devoted to demilitarization specified a date of February 15, 1989, for the withdrawal of Soviet troops. The Soviet-installed government in Afghanistan, headed by Muhammad Najibullah, who had led the country since 1986, sought to maintain rule after the expulsion of Soviet troops. As the Soviets withdrew, the seven leading parties of the Mujahedeen opposition banned together to form the Afghan Interim Government and displace the Najibullah government. It was three years before the alliance captured Kabul in April 1992. During the conflict, political control was fragmented among regionally and ethnically based war-era military chieftains. Ideological and ethnic divisions among the leaders of the wartime coalition reemerged after the fall of Kabul, as the commanders turned on one another in a contest for ultimate control of the government. The contest quickly turned back to the instruments of war.

In the wake of the disastrous fighting among the Afghanistan factions, a new force emerged. In 1994 a coalition of Mujahedeen forces calling itself the Taliban took up arms to cleanse the country of the corruption and internal strife of the post-conflict competition among warlords. The Taliban was born as a movement in the northwestern province of Pakistan during the early 1990s. Capturing Kandahar in November 1994, and Herat in September 1995, Taliban fighters pushed through southern and eastern Afghanistan, uniting the primarily ethnic Pashtu areas under strict Islamic rule. The Taliban succeeded in capturing Kabul in September 1996, ousting the government of Burhanuddin Rabbani and establishing Taliban supreme leader Mullah Mohammad Omar in power. Resistance to the Taliban remained active, however, particularly in the northern and eastern part of the country. Commander Abdul Rashid Dostam's "National Islamic Movement," consisting heavily of ethnic Uzbeks, controlled several north-central provinces. Commander Ahmad Shah Masood controlled the ethnic Tajik majority areas of the northeast.[12]

Given the illegitimate conditions under which the Taliban gained power and established political control, the UN and other international organizations were unwilling to recognize the Taliban as the legitimate government of Afghanistan, deferring decision on diplomatic credentials. The Organization of the Islamic Conference left the Afghan seat vacant until the question of legitimacy could be resolved. Major world governments and international organizations instituted sanctions against the Taliban. Osama bin Laden had been allied with some of the most ruthless leaders of the Mujahedeen Afghan resistance, for instance, Gulbuddin Hekmatyar, the head of the Islamic Party. After the fall of the Soviet-backed government, Hekmatyar briefly became Afghanistan's prime minister and sought to capture Kabul in an effort to dislodge more moderate Mujahedeen leaders.

After the Soviet withdrawal from Afghanistan, Moscow and Washington were anxious to redirect their attentions to more pressing issues closer to home. But the shadow of Afghanistan continued to loom in Central Asia. In Afghanistan's vacuum of legitimate political control, a coalition of professional terrorists, drug traffickers, and Muslim zealots combined resources to take control of the country. The situation grew increasingly severe as formal training camps were established to train terrorists recruited throughout the Middle East. Afghanistan became the world's number one producer of opium.[13] The "frontline" states bordering Afghanistan sought international assistance to resolve the drug trafficking and terrorism problems, but no major outside government was anxious to engage in solving Afghanistan's problems, given the country's immediate history. After terrorist bomb attacks on two U.S. embassies in Africa in 1998, the U.S. began to mobilize against the dangers of terrorism.

While the Eurasian governments have been unsuccessful in uniting to cooperate to achieve many regional economic and security objectives, they found common cause in opposition to the Taliban's northward expansion. The Russian government saw the Afghanistan-Tajikistan border as a porous integument through which terrorism, weapons, and drugs penetrated en route to destinations in Eurasia. At the same time, the border was seen as a cultural divide between the values of western civilization and the lawlessness of Afghanistan. For this reason, Tajikistan President Imomali Rahmonov in November 1999 proudly observed that Tajikistan considered Russia to be its "strategic partner."[14] Without interruption since the disintegration of the USSR, Russian and Tajik military units have policed the Tajikistan border with

Afghanistan. In December 1999, then Prime Minister Putin described Uzbekistan as "Russia's strategic partner."[15] In June 2000 Kyrgyzstan President Askar Akaev announced that "Russia always has been and will remain Kyrgyzstan's principal strategic partner."[16] The Taliban has also given rise to other forms of cooperation. The Shanghai Organization Countries (China, Kazakhstan, Kyrgyzstan, Russia, Tajikistan, and Uzbekistan) present a united front in their assessment that the Taliban-inspired insurrectionism constitutes one of the chief challenges to political normalization in the Central Asian region.

As it grew apparent that the bombings of the U.S. embassies in Africa were the work of terrorist leader Osama bin Laden, a revolutionary from a wealthy Saudi Arabian family who was also an early financial supporter of the Taliban and who has been given sanctuary by the regime, the U.S. took another tack. On August 20, 1998, U.S. jets armed with cruise missiles attacked a terrorist camp under the direction of the exiled Saudi terrorist chieftain. The U.S. teamed up with its former adversary, Russia, to jointly sponsor a UN Security Council Resolution in an effort to stem the flow of drugs, stop the training of terrorists, and demand that Afghanistan surrender Osama bin Laden.[17] When the original sanctions failed to gain the Taliban's compliance, the Security Council adopted a second round of yet more severe sanctions in December 2000.[18]

Soon after the terrorist attack on the U.S., the Bush administration started a diplomatic initiative followed by military operations aimed at achieving the goals of the UN security resolutions. The U.S. secured a commitment from Pakistani Prime Minister Pervez Musharraf to provide logistical support for the strike capabilities of American forces. Soon afterward leaders in Uzbekistan, Tajikistan, Turkmenistan, and Kyrgyzstan agreed to assist in the anti-Taliban operations and, by implication, to close diplomatic relationships with the U.S. The U.S. rationale for the anti-Taliban war was that it was an issue of international security and justice. But some observers saw that beneath these issues was the competition over control of the region's resources, and this, above all, meant control over the region's energy resources.[19]

The Politics of Energy in the Caspian Basin

The Central Asian countries are highly interdependent with respect to energy resources, transportation infrastructure, and markets.[20] The greatest source of wealth in the region is natural resources, particularly gas and oil. The region's major rich oil and gas reserves are located in and around the world's largest inland body of water, the Caspian Sea. Five countries border the Caspian Sea. Much of the oil wealth is located under the Caspian Sea itself or in areas adjoining the sea in Azerbaijan, Kazakhstan, Turkmenistan, and Uzbekistan. The potential for increasing oil and gas production in the region is great. Oil industry analysts expect that the region could be exporting as much as 2 million barrels a day by 2010. But because all the region's oil-producing countries are landlocked, routes to the market invariably involve shipment through third-party countries. The previously existing export routes for Caspian energy were northward, running through Russia to Europe. As a result, Russia had commercial and political interests in retaining its position as the conduit for the Caspian region's energy resources. But Russia's monopoly position produced anxieties among Caspian producers and shippers with respect to reliability, security, transparency, equitable tariffs, and access. The region's geography and the differing national interests of the countries make access to market a matter of mutual agreement, or at least mutual negotiation. There is no single good route for shipping out oil and gas. There is a wide spectrum of geographical impediments and political risks associated with any transport route from the Caspian. Limited capacity in the existing old Soviet routes across Russia requires that this pipeline network be upgraded. Shipping oil to Turkey through routes that cross Armenia and Georgia require that entirely new pipelines be constructed. Moreover, there is opposition to a trans-Turkey route from influential constituencies in Turkey who want to limit traffic through the Bosporus for both environmental and commercial reasons. The existing and projected Transcaucasian pipeline routes run through regions with histories of separatism and insurgency, including Chechnya, Abkhazia, and Kurdistan. Sanctions imposed on Iran by the U.S. for sponsorship of terrorism preclude U.S. oil companies from participating in routes that transverse Iran.[21]

The geographical and political impediments to development had an effect that tended to separate rather than align the countries. Different countries had different responses to the situation. Kazakhstan, for instance, sought multiple routes but found itself, at least initially, highly reliant upon Russian transportation and refining capacity. When the USSR came to an end, Kazakhstan took claim to the natural resources on its territory, including the oil and gas reserves. Much of Kazakhstan's oil wealth was located in the northwestern section of the country, close to the southern Russia border. In 1991 the Kazakhstan government began negotiating what the Kazakhstan press referred to at the time as the "the deal of the century" to develop Kazakhstan's "Tengiz" oil fields. The Kazakhstan government joined the large multinational oil firm Chevron to form a joint venture, "Tengizchevroil." The agreement committed Chevron to spend $20 billion over twenty years to develop the Tengiz field's 6 billion barrels of proven reserves. Other firms, including Birlesmis Muhendisler Burosu, British Gas, and Agip, also committed investment in the country's oil fields. Russia first sought to impose high taxes and surcharges on the movement of gas and oil and then, in order to block economic development that might run contrary to the interest of Russian energy firms, sought to blockade its southern neighbors by cutting off access to foreign markets entirely.

Chevron's returns from the Tengizchevroil joint venture, falling far short of expectations because of bureaucratic red tape, were threatened by Russia's intransigence. When the Tengizchevoil agreement was first reached, it was anticipated that eventually Kazakhstan would be exporting as much as 700,000 barrels a day. But as late as October 1995, the Tengiz field was producing only 50,000 barrels a day, about half of the expected pumping capacity. Chevron and the leadership of the Kazakhstan government began seeking to enlarge the players in the Central Asian energy market to increase the political pressure on Russia.

Another example of the divisive effects of the energy competition is illustrated by Turkmenistan's energy policy. Turkmenistan had been supplying gas to the Soviet gas distribution system at artificially low prices. At independence, Turkmenistan quickly sought to turn its gas reserves to its financial advantage. Soon after independence, Turkmenistan announced that it would sell gas only at market prices. Gas deliveries to Armenia, Georgia, Ukraine, and Russia were suspended. The Turkmenistan government sought to recover what it said were debts for gas sales from Kazakhstan, Kyrgyzstan, and Tajikistan.[22] A sharp dispute erupted with one of Turkmenistan's largest clients, Ukraine. Ukrainian foreign minister Vitalii Fokin charged that Turkmenistan's behavior was "uncivilized" and amounted to "economic blackmail."[23] Turkmenistan prevailed; Ukraine settled for the Turkmen terms. But Turkmen gas was still reliant on the existing gas transportation system. Turkmen gas revenues continued to decline through the 1990s.

As a result of these disputes, the Russian Foreign Ministry increasingly began to see the competition over gas routes as the new "Great Game." Spurred on by the nominally privatized oil companies with close links to the government through interlocking directorates and financial influence, Russian government officials came to see the goals as regaining control over the Caspian's oil and gas resources and keeping U.S. companies out unless it was on Russia's terms. Russians accused Azerbaijan and Kazakhstan of acting independently to exploit natural resources in the Caspian basin, oil resources that the Russians claimed should belong to all littoral states. The disagreements forced negotiators to the table to develop an overall regional development strategy. These negotiations resulted in the signing of a "Caspian Littoral Agreement." The agreement included Azerbaijan, Iran, Kazakhstan, Russia, and Turkmenistan. The agreement was designed to coordinate trade routes, regulate the access to natural and mineral resources, and unite efforts at environmental protection. The agreement also established a "Caspian Council," consisting of a secretariat and four specialist committees. The council was to be politically controlled by an intergovernmental council representing the littoral states.

The southern-tier countries began discussing ways to bring energy supplies to market without having to pass through Russian territory. Under the auspices of the Iranian-led Economic Cooperation Organization (ECO), the Central Asian countries discussed four major pipeline projects: Baku-Ceyhan (Turkey); Kazakhstan-Turkmenistan-Afghanistan-Pakistan to the Indian Ocean; Turkmenistan-Iran-Turkey toward Europe; and the supply of Uzbek and Turkmen gas to Pakistan through Afghanistan.

Meanwhile, the U.S. was seeking to facilitate infrastructure development that would serve other American foreign policy aims. In January 1995, the U.S. government adopted the idea of promoting "the Turkish route" for Caspian oil and gas, arguing that this was essential for regional security, economic independence, and commercial development. In 1998 the Clinton administration unveiled its Caspian Basin Initiative, with the creation of the post of special advisor and establishment of an Ankara-based Caspian Finance Center to coordinate efforts by its export finance agencies in the region. In October 1995, a partial solution to the problem of transporting oil from Azerbaijan was found with the backing of the U.S. government. Two routes were chosen for early oil production from Azerbaijan International Operating Company (AIOC) projects in Azerbaijan: one stretching from Baku through Chechnya to Russia's Black Sea port of Novorossiisk and a second reaching from Baku across Georgia to its port of Supsa on the Black Sea.

The Turkmen gas pipeline "Eurasian Energy Corridor" resulted in a joint venture by Bechtel and General Electric, later joined by Royal Dutch/Shell to put together a package for the construction of the Trans-Caspian Pipeline (TCP) that would carry up to 30 billion cubic meters (bcm) per year from Turkmenistan to Azerbaijan, along the Caspian seabed on a 1,250-mile route from Turkmenistan through Azerbaijan and Georgia to Turkey. Negotiations broke down in a dispute between Turkmen officials and their Azeribaijan counterparts, who wanted to export their own gas from the newly discovered Shah Deniz field to ship to Turkey.

At the same time the TCP was faltering, a Russian-Italian-Turkish joint venture dubbed "Blue Stream," a 750-mile-long gas pipeline from Russia to Turkey beneath the Black Sea, went forward despite substantial technical and political difficulties. Completion of the Itwin 24-inch natural gas pipelines at depths reaching nearly 7,000 feet took place in summer 2002. In October 2002, Russia's Gazprom, the co-owner/operator of the pipeline, announced that Blue Stream would begin carrying processed gas from Djubga, Russia, across the Black Sea to Samsun, Turkey, starting in December 2002. At full capacity, Blue Stream is designed to carry 565 billion cubic feet of natural gas per year from Russia to Turkey for distribution in western markets.

The Eurasian Integration Movement and "Soft Hegemony"

The shifting fortunes in the competition for advantage in oil and gas routes were only the most visible manifestation of a more fundamental competition among the Eurasian states. Competition among the former Soviet Eurasian states continued during the first decade to defeat cooperation. Persistent one-upmanship among the countries appeared to make sustained interstate cooperation impossible. Faced with shirking and free-riding, some countries sought to break out of the CIS multilateral framework to go it alone by forming bilateral and regional agreements. Kazakhstan's president Nursultan Nazarbaev was one of the first CIS leaders to openly criticize the failures of the CIS mechanisms to produce sustainable cooperative results. Nazarbaev observed as early as 1994,

> Since the time of the establishment of the Commonwealth of Independent States, roughly 400 agreements have been adopted. However, as yet there have been no substantive results because individual national governments continue to reject certain provisions and interpret the meaning of the agreements in their own interest.[24]

Treaties were adopted and laws were put on the books, but the anticipated level of cooperation did not materialize.

To overcome the problems of interstate cooperation, Kazakhstan's President Nazarbaev announced his plan for a "Eurasian Union" in 1994.[25] Nazarbaev foresaw an interstate union that would realize the idea of creating a single economic, policy, and information space throughout Eurasia. Nazarbaev's concept was modeled on other successful regional integration efforts, in particular, the European Economic Community (which has since become known as the European Union, or EU), the North American Free Trade Agreement (NAFTA), the Association of South East Asian Nations, (ASEAN), and the Common Market of the Southern Cone (known as Mercosur).

For several years Russian officials were indifferent toward Nursultan Nazarbaev's proposals for the establishment of a "Eurasian Union." Russian policy officials during the Yeltsin years viewed Nazarbaev's proposals as an unnecessary complement to the CIS. The continuing unwillingness of key CIS members, particularly Ukraine and Uzbekistan, to acquiescence to Russian dominance in the CIS, however, made it difficult for the Russians to realize what they regarded as the promise of the CIS. During the 1990s, Russian policy toward the Central Asian states grew increasingly fragmented. By the time Yeltsin left office, a consensus had already formed in the Ministry of Foreign Affairs that some new approach toward Central Asia was necessary. In spring 2000, Russian officials abruptly shifted their position on Nazarbaev's proposals. The Russians accepted Nazarbaev's ideas and even began energetically negotiating the expansion of Nazarbaev's original proposal. In October 2000, the presidents of five states gathered in Astana to sign the foundation documents creating a Eurasian Economic Community (*Evraziskoe Ekonomicheskoe Soobshchestvo*), the EvrAzES. In May 2001, EvrAzES was formally brought into being following ratification of the treaty by the countries' parliaments. By autumn 2001, EvrAzES headquarters was functioning in Moscow under the direction of Grigorii Rapota, former director of *Rossvooruzhenie*, the Russian state-controlled arms-trading enterprise.[26]

At this point Ukraine and Uzbekistan were still holdouts. Previously Ukrainians regarded the EvrAzES, in the words of one parliamentary deputy, as simply a "second edition of the USSR."[27] But cooperation with Russia was becoming as much an economic issue as a political issue. Ukraine has a massive debt burden to the state-controlled Russian gas monopoly, Gazprom. Meeting with Vladimir Putin in Sochi on May 18, 2002, Leonid Kuchma said that he would not rule out Ukrainian membership in the EvrAzES.[28] Russian PM Mikhail Kasyanov floated the idea that Ukraine would be a welcome member.[29] The announcement in April 2002 that Ukraine would indeed participate as an observer in the EvrAzES changed the situation entirely.[30] This move was immediately applauded by the speaker of the Russian Duma, Gennadi Selenev, in a way that suggested that Ukraine would be welcomed into full participation.[31] Thus, by late 2001, the EvrAzES emerged as Russia's principal means of reasserting its influence in the south Eurasian region, not through heavy-handed coercion but through "soft hegemony."

Conclusion: After Afghanistan

Along its southern frontiers, Russia has long shared cultural and political borders with the world of Islam. As the disintegration of the USSR gave expression to long-suppressed nationalist and sectarian tendencies in the region, it simultaneously led to a climate of indeterminacy and lawlessness that offered fertile ground to the interplay among militant Muslim separatists, insurgents, and terrorists in Chechnya, Dagestan, and other Russian border regions. The links, both ideological and material, among the militant movements of the Caucasus, Afghanistan, and the Middle East, gave credence to the impression that a clash of civilizations was becoming the core value of Russia's southern foreign policy.[32] But while it is true that there are elements of religious fervor in this competition over the hearts and minds of Eurasia, it is easy to read too much determinism into this struggle. Many of south Eurasia's warriors are less "Warriors of God" than they are "Warriors of War." As Sergei Mironov, the speaker of the Russian Federation Council, has argued, "the roots of terrorism in Afghanistan, Chechnya, and the Middle East are less ideological in nature than they are financial."[33] Muslims and non-Muslims have lived in Eurasia for centuries side-by-side. At bottom, Russian security strategy in Eurasia does not comprehend the "Great Game" in terms of a confrontation of cultures, but in terms of competing interests vying for influence and control in the area of Eurasia to the south of Russia's borders.

The terrorist attack on the New York Trade Towers and the Pentagon transformed American politics. The terrorist attack just as clearly transformed interstate relations in Eurasia. As soon as a commitment was made to oust the Taliban from control of Afghanistan, the U.S. government recognized that the successful prosecution of anti-Taliban operations in Afghanistan would require close U.S. cooperation with Central Asian frontline states. This implied a new level of U.S.-Russian cooperation with respect to Central Asia. Previous U.S.-Russian cooperation on UN Security Council Resolutions 1267 and 1333, which imposed sanctions on the Taliban, had not been successful in winning the extradition of Osama bin Laden or ending Taliban support for terrorism. But the experience was successful in demonstrating that Russia and the U.S., once bitter antagonists over Afghanistan, could work collaboratively for mutual advantage in the region. The expression of support from President Putin immediately after the September 11 events reassured the U.S. administration that further cooperation in the Eurasian region was possible. For their own part, the Central Asian countries, anxious to rid themselves of the menace of insurgency and narco-business that had been emanating from Afghanistan for the previous several years, were happy to lend enthusiastic support to the U.S. operations to oust the Taliban.[34]

Initially the prospect of a direct U.S. presence, particularly a military presence, in Central Asia caused concern with Russian nationalist circles who viewed American actions as simply a manifestation of American global hegemonic aspirations—globalization "po-amerikanski." But the Russian government leaders dismissed this criticism. Step-by-step, Russia acceded to the

U.S.'s growing military presence in the region. The stationing of U.S. troops at Khanabad airbase in Uzbekistan and Manas airbase in Kyrgyzstan in October 2001, the use of military facilities in Tajikistan and Kazakhstan, and the cross-border transit agreements with Turkmenistan and Uzbekistan in November 2001 raised no objections from Moscow. Russian acquiescence to the decision in March 2002 to send U.S. advisors to train antiterrorist troops in Georgia's Pankisi Gorge crossed yet another major psychological barrier. By March 2002, the U.S. had a low-profile military presence throughout the entire southern Eurasian region. Finally the adoption of the Rome Declaration on May 28, 2002, establishing the NATO-Russia Council, brought the process full circle.[35] Russia had become a junior member in the very alliance that was formed to combat the USSR and had willingly accepted U.S. security cooperation throughout its southern Eurasian sphere of influence. These steps constituted a major U.S.-Russian entente.

The U.S.-Russian entente had profound implications for Russia's evolving strategy in the post-Afghanistan strategic readjustment of Eurasia. For one thing, Russian policy makers are aware that Russia is important to other countries, as Robert Legvold has expressed it, primarily for three reasons: "the atom, the veto, and the location."[36] Russia has used its Soviet-era nuclear and ballistic missile capability as a trump card, never in sight but never out of mind. Putin's pragmatism allowed him to be flexible on tactics while striving to achieve larger goals. On December 13, 2001, Moscow acquiesced to U.S. withdrawal from the 1972 ABM Treaty. On May 24, 2002, the Russians signed the Moscow Treaty on Strategic Offensive Reductions, agreeing to reductions in strategic arsenals. In exchange for flexibility, Russia was rewarded with the long-sought-after status of a market economy, relieving Russian steel and titanium producers from the most onerous provisions of the antidumping legislation.[37]

Another major implication of the new entente was that Russia's closer relationship with the U.S. would take place at the expense of Russia's commercially valuable relations with its southern neighbors Iran and Iraq. Iraq, a major oil and gas producer, has offered Russian companies the rights to vast future energy development projects valued at as much as $50 billion. Russia has maintained the largest share of Iraqi foreign trade under the United Nations oil-for-food program. Russia was helping Iran construct a 1,000-megawatt commercial power nuclear reactor in southern Iran at Bushehr. The Bushehr facility includes a uranium processing capability that could be modified to produce fissile material for nuclear weapons. Russian technicians are also helping Iran develop the Shahab-4 missile with a range of 1,250 miles. The Russian Ministry of Atomic Energy and *Rosoboronexport* lobbied hard to get these projects. Moscow diplomatic officials had argued that adequate safeguards were in place for the nuclear reactor in Bushehr and that the ballistic missile technology was not a threat to regional stability. But the new entente would mean that Moscow would have to weigh the costs for continuing these relationships with its client states.

Another likely consequence of the U.S.-Russian entente is that it will lead Russia to put greater emphasis on Russia's policy of "soft hegemony." It is true that at the most fundamental psychological level there is a still strong feeling within the Russian policy-making community that south Eurasia is Russia's "backyard" and that the countries of the region should have a special, exclusive relationship with Moscow. Russian national pride was badly stung by the circumstances of the disintegration of the USSR. Moscow exchanged superpower status for that of a lower-middle-income country and got little in exchange except for the relaxation in international tensions and an opportunity to abandon the failed Marxist experiment. While that is a relief, it is not much of a consolation. Russia's support for the idea of a "North-South" infrastructure could help to position Russia to capitalize on the new links between Asia and Europe. Pakistan, Turkmenistan, and Afghanistan leaders were anxious to formalize their resolve to link south Asia with Eurasia in May 2002 when they signed a joint agreement to build a gas pipeline from Central Asia, through Afghanistan, and on to Pakistan.[38] The large oil consumers, with the U.S. at the forefront, seem unable to reduce consumption through conservation and other means and thus continue to be vulnerable to oil supply shocks. Energy security thus implies greater security of supply. Arab resentment of America's support for Israel and threats against Iraq, the frustrations of many younger Arabs with their own lack of freedom and jobs, and the domestic instability within the Saudi kingdom suggest that Russia and the Caspian region could be critical in times of emergency. Russia's participation in such projects as the trans-Eurasian Afghanistan pipeline, the Blue Stream, the Caspian Pipeline Consortium, and other infrastructure projects will prevent Russia from being bypassed in the race to link the infrastructures of Europe and Asia. Russia's enthusiasm for the Eurasian economic integration efforts is part of a larger design to exploit Russia's pivotal location between Asia and Europe in an effort to recapture the benefits of hegemony without paying the price of being a superpower.

Notes

1. See William Fierman, *Soviet Central Asia: The Failed Transformation* (Boulder: Westview Press, 1991); Boris Rumer, *Soviet Central Asia: A Tragic Experiment* (Boston: Unwin Hyman, 1989); and Gregory Massell, *The Surrogate Proletariat* (Princeton: Princeton University Press, 1975).

2. This is a major theme in the Soviet economic literature of the 1970s and 1980s

by Boris Rumer. See Boris Rumer, *Soviet Central Asia: A Tragic Experiment.*

3. Gregory Gleason, "The Federal Formula and the Collapse of the USSR," *Publius: The Journal of Federalism*, vol. 22 (Summer 1992), pp. 141–163.

4. Opposing complete abandonment of Russia's interests in Central Asia, former Russian Foreign Minister Andrei Kozyrev argued forcefully that the Central Asian areas should remain under Russian "protection." Andrei Kozyrev, "Chego khochet Rossiia v Tadzhikistane," *Izvestia*, August 4, 1993, p. 4.

5. Jeff Chinn Robert Kaiser, *Russians as the New Minority* (Boulder: Westview Press, 1996).

6. Richard Sakwa and Mark Webber, "The Commonwealth of Independent States, 1991–1998: Stagnation and Survival," *Europe-Asia Studies*, vol. 51, no. 3 (1999), pp. 379–415.

7. Gregory Gleason "Inter-state Cooperation in Central Asia from the CIS to the Shanghai Forum," *Europe-Asia Studies*, vol. 53, no. 7 (2001), pp. 1077–1095.

8. Valerie Bunce, "The Political Economy of Postsocialism," *Slavic Review*, vol. 58, no. 4 (Winter 1999), pp. 756–793.

9. Anders Aslund, *Building Capitalism: The Transformation of the Former Soviet Bloc* (Cambridge: Cambridge University Press, 2001).

10. Russia's "New Foreign Policy Concept" was approved by the Russian Security Council on March 24, 2000, and approved by President Putin on July 2, 2000. The Concept was printed in *Rossiskaia gazeta*, July 11, 2000. The Concept is available on the website of the Russian Foreign Ministry: www.In.mid.ru/website/ns-osndoc.nsf/ osnvnpol. The Concept serves as a guidance document for the Ministry of Foreign Affairs, articulating the goals and objectives of Russian foreign policy. The preceding programmatic statement, often referred to as the "Kozyrev doctrine," was adopted in April 1993 during Andrei Kozyrev's tenure as Minister of Foreign Affairs.

11. Zahir Shah was established as monarch of Afghanistan in 1933. His forty years of rule came to an end in July 1973 when his previous premier, Muhammad Daoud, overthrew the monarchy in a military coup. Five years later, in April 1978, Daoud was overthrown in a coup organized by the Marxist People's Democratic Party of Afghanistan (PDPA). Although Marxist in orientation, the PDPA was anxious to move out of the sphere of influence of Moscow. After the assassination of the PDPA's leader, Nur Muhammad Taraki, in September 1979, the Soviet Union, concerned about the turn of events in neighboring Iran and the possible loss of an allied country on its sensitive southern border, responded by invading Afghanistan to establish a puppet government.

12. Ahmed Shah Masood was the leader of the Jumbesh-e Melli-ye Islami-ye Afghanistan, the primarily ethnic Tajik resistance based in the northeastern areas of Afghanistan. Masood was assassinated in September 2001.

13. UNDCCP, "Afghanistan Opium Cultivation in 2000 Substantially Unchanged: Country Still the Largest Opium Producer in the World," UNIS/NAR/696 (September 15, 2000). Available at http://www.odccp.org/press_release_2000–09–15_1.html.

14. *RFE/RL Newsline*, vol. 3, no. 224, November 17, 1999. Available at http://www.rferl.org/newsline/1999/11/171199.asp.

15. *RFE/RL Newsline* vol. 3, no. 240, December 13, 1999. Available at http://www.rferl.org/newsline/1999/12/131299.asp.

16. *RFE/RL Newsline*, vol. 4, no. 128, July 3, 2000. Available at http://www.rferl.org/newsline/2000/07/030700.asp.

17. UN Security Council Resolution 1267 of October 15, 1999. Available at www.un.org/Docs/scres/1999/99sc1267.htm.

18. UN Security Council Resolution 1333 of December 19, 2000. Available at www.un.org/Docs/scres/2000/res1333e.pdf.

19. Ahmed Rashid, *Taliban: Militant Islam, Oil and Fundamentalism in Central Asia* (New Haven: Yale University Press, 2001).

20. Douglas W. Blum, "Domestic Politics and Russia's Caspian Policy," *Post-Soviet Affairs*, vol. 14, no. 2 (April-June 1998), pp. 137–164.

21. The Iran-Libya Sanctions Act (ILSA) was passed by the U.S. Congress and signed into law in August 1996. ILSA imposes mandatory and discretionary sanctions on non-U.S. companies that invest more than $20 million annually (lowered in August 1997 from $40 million) in the Iranian oil and gas sectors. U.S. Executive Order 13059, signed in August 1997, prohibited virtually all trade and investment activities by U.S. citizens in Iran.

22. *Nezavisimaia gazeta*, May 8, 1992, p. 1.

23. *Izvestia*, March 5, 1992, p. 1.

24. Nursultan Nazarbaev, *Five Years of Independence* (Almaty: Kazakstan, 1996), p. 234.

25. Nursultan Nazarbaev announced in September 1994 that an executive office for the "Eurasian Commonwealth" would be established in Almaty. Nazarbaev opined that he had from the very earliest days of the CIS argued in favor of such an organization, but, he implied, other CIS leaders were reluctant to take this step. Nursultan Nazarbaev, "V poiske novoi integratsii," in *Piat' let nezavisimosti* (Almaty: Ilim, 1996), pp. 233–242.

26. Rapota had prior to this served as a section chief in the KGB First Main Directorate.

27. Ukrainian Minister of Foreign Affairs Anatolii Zlenko and Secretary of State Aleksandr Chalyi were quoted as saying that Ukraine could not simultaneously integrate in the western structures and move in the direction of cooperation with the EvrAzES. "Ukraine has chosen its vector—it is European." Tatiana Ivzhenko, "Ukrainu zagnali "za flazhki," *Nezavisimaia gazeta*, May 17, 2002, p. 1.

28. Armen Khanbabian, "NATO, Gaz i khoroshaia pogoda," *Nezavisimaia gazeta*, May 20, 2002, p. 1.

29. "Kasyanov pozval Ukrainu v EvrAzES." *Kabar*, April 11, 2002. Available at http://www.kabar.kg/02/Apr/11/40.html.

30. "Ukraine will Assume Observer Role in the EvrAzES," *RIA Novosti*, May 14, 2002. Available at www.ukraine.ru/news/134027.html.

31. "Seleznev Welcomes the Decision of the Presidents of Ukraine and Moldova to Send Observers to EvrAzES," *RIA Novosti*, May 24, 2002. Available at www.e-journal.ru/n25036904.html.

32. Samuel P. Huntington, *The Clash of Civilizations and the Remaking of World Order* (New York: Touchstone, 1998).

33. "Russia Denies Shift in Mideast Policy," *Vesti RTR*, March 13, 2002. Available at www.rtr-vesti.ru/news.html?id=729&tid=503&date=11–03–2002.

34. This was accompanied by a financial inducement as well. U.S. assistance to Uzbekistan, including the fall 2001 supplemental budget and the regular budget, rose from $55.9 million in fiscal year 2001 to $161.8 million in fiscal year 2002.

35. While not a formal member of NATO, the agreement welcomes Russia to the European community. Russia is not a member nor bound by the provisions of the collective security arrangement that requires members to come to the defense of the other members. Russia has no veto power and no vote in the consideration of new

members. Representatives from Moscow first took part in meetings at NATO in 1991, as part of the North Atlantic Cooperation Council (NACC). In 1997, the NATO-Russia "Founding Act" established a NATO-Russia Permanent Joint Council (PJC). The PJC held its last meeting in Reykjavik on May 14, 2002.

36. Robert Legvold, "Russia's Unformed Foreign Policy," *Foreign Affairs*, vol. 80, no. 5 (September–October 2001), pp. 62–75.

37. The U.S. Department of Commerce announced the decision to reclassify Russia as a market economy on June 6, 2002.

38. On May 30, 2002, in Islambad, Khamid Karzai, Saparmurat Niyazov, and Pervez Musharraf signed an agreement to build a 1,500-km-long gas pipeline from Turkmenistan through Afghanistan to the Pakistan port of Gvadar.

For Further Reading

Peter Hopkirk. *The Great Game: The Struggle for Empire in Central Asia* (New York: Kodansha International, 1990).

Samuel P. Huntington. *The Clash of Civilizations and the Remaking of World Order* (New York: Penguin, 1997).

Nursultan Nazarbaev. *Kazakhstan-2030: Prosperity, Security and Ever Growing Welfare of all the Kazakhstanis* (Almaty: Ylim, 1997).

Islam Karimov. *Uzbekistan on the Threshold of the Twenty-First Century* (London: St. Martin's Press, 1998).

Zbigniew Brzezinski. *The Grand Chessboard: American Primacy and Its Geostrategic Imperatives* (New York: Basic Books, 1998).

Olivier Roy. *The New Central Asia: The Creation of Nations* (New York: New York University Press, 2000).

Anders Åslund. *Building Capitalism: The Transformation of the Former Soviet Bloc* (Cambridge: Cambridge University Press, 2001).

Pauline Jones Luong. *Institutional Changes and Political Continuity in Post-Soviet Central Asia: Power, Perceptions, and Pacts* (Cambridge: Cambridge University Press, 2002).

Annual Energy Outlook 2002. U.S. Department of Energy (Washington, DC: U.S. Government Printing Office, 2002).

Martha Brill Olcott. *Kazakhstan: Unfulfilled Promise* (Washington, DC: Carnegie Endowment for International Peace, 2002).

5

Russian-American Relations

Herbert J. Ellison

The development of a foreign and security policy has been one of the most difficult tasks confronting the leadership for the new Russian state since its creation in 1991, and no part of the task has been more challenging than the Russian relationship with the United States. It has involved a vast transformation both of policy and of mutual perceptions and expectations developed over a half-century of political and ideological confrontation.

Russian-American relations since the formation of the new Russian state in 1991 can be divided into three phases. The first phase (1991–95), led by the young Foreign Minister Andrei Kozyrev, was notable for an optimistic pursuit of partnership with the U.S. and its allies. The policy met discouragement both from the lack of support for Russia's difficult economic transition and from the sense of Russian exclusion from the American-dominated, post-communist security systems in Europe and East Asia. The spectacular electoral victory of the Communists and Nationalists, Kozyrev's fiercest critics, in the 1993 and 1995 Duma led Yeltsin to dismiss him as a political liability as he confronted a second presidential election with the polls favoring his Communist competitor, Gennadii Zyuganov.

Kozyrev's successor, Yevgenii Primakov, who dominated foreign policy for the next four years (1996–99), brought a major policy change. He portrayed America not as a likely Russian partner but as a dangerous rival, the sole surviving superpower busily pursuing global hegemony in a "unipolar" world. During his policy leadership, Russia pursued a "strategic partnership" with China and cooperation with other states to build a separate "pole" of power to balance and check that of the U.S. His views on foreign policy were dominant during his tenure as foreign minister (1996–1998) and as prime minister (1998–1999).

The third phase of Russian post-communist policy began with the Putin era (2000–). Official policy pronouncements concerning the U.S. at first reflected the language and concepts of the Primakov confrontation with the

U.S., which had reached a climax in the 1999 NATO action against Yugoslavia and the debate over U.S. plans for nuclear missile defense beginning in the same year. The relationship began to improve with the advent of the Bush leadership, and underwent spectacular change following the events of September 11, 2001. From the opening of his presidency, Putin had made it clear that his primary focus on Russian economic expansion required close cooperation with Europe and the U.S. His remarkable support of the U.S. and NATO after the terrorist attack, and a new American and European receptiveness to security cooperation with Russia, brought a new Russian approach to relations with the U.S. more like that of the early Yeltsin years, strengthened by the growing evidence both of Russia's growing political and economic strength and of its potential role in Western economic and political security.

Phase I: Foreign Policy Revolution (1991–1995)

In foreign as in domestic policy, Boris Yeltsin proved himself a revolutionary. Certainly the way had been prepared by the sweeping policy changes of Gorbachev and Shevardnadze, but Yeltsin's early proposals went much further, calling for full partnership with the U.S. and the other democratic states, for a cooperative program aimed at complete elimination of weapons of mass destruction, and for a large expansion of the role of the United Nations in the management of international conflict.

Yeltsin lost no time in seeking to find common ground with the Western states on arms control. On January 29, 1992, little more than a month after the creation of the new Russia, he issued a comprehensive statement "On Russia's Policy in the Field of Arms Limitation and Reduction."[1] Beginning with the assertion that "Nuclear weapons and other means of mass destruction in the world must be eliminated," he announced that the new Russian leadership would recognize all arms commitments of its predecessor, and would also pursue a "radical reduction in nuclear arms." He proposed the formation of an "International Agency for the Reduction of Nuclear Arms," affirmed his commitment to the START treaty—already submitted to the Russian Supreme Soviet—and announced a series of unilateral measures undertaken by Russia to reduce its weapons arsenal.

The statement was comprehensive in its coverage: tactical and strategic nuclear weapons; antimissile defense and space; nuclear weapons testing and production of fissionable materials for weapons; and nonproliferation of weapons of mass destruction, including chemical and biological weapons. It also proposed dramatic reductions in the Russian defense budget and conversion of military industries to civilian production needs.

Yeltsin stressed Russia's reaffirmation of its commitments under the Treaty on the Nonproliferation of Nuclear Weapons, and announced that "we expect the earliest possible accession to the treaty, as non-nuclear-weapon states, of Belarus, Kazakhstan and Ukraine, as well as of other CIS member-states." The three states specifically mentioned, which contained huge arsenals of nuclear weapons, were expected to transfer them to Russian control.

He also announced preparations for Russian acceptance of the guarantees of the International Atomic Energy Agency on nuclear exports, and endorsed the international system of nonproliferation of missiles and missile technology and the efforts of the "Australian group" to monitor chemical exports. He noted the work in the Russian Federation on legislation to regulate exports of "dual-purpose materials, equipment and technologies that could be used to create nuclear, chemical or biological weapons, as well as combat missiles," and talked of planning for a monitoring system and of acceptance of the "guiding principles for trade in weapons that were approved in London in October 1991."

Shortly after his remarkable statement, Yeltsin departed for New York and a speech to the UN Security Council where he spoke on January 31, 1992. Though most of the speech duplicated his earlier statement in Moscow, it contained a proposal on missile defense that was highly original:

> I believe that the time has come to pose the question of creating a global system for the protection of the world community. It could be based on a reorientation of the U.S. Strategic Defense Initiative (SDI) using high technologies developed in Russia's defense complex.[2]

In a press interview on the following day, he amplified his proposal by suggesting, "this system could be manufactured and put into place through the joint efforts of the U.S. and Russia, and perhaps other nuclear powers as well."[3] The issue of nuclear missile defense he had raised received no response from the American side, but would reemerge in the 1990s as a major point of friction in U.S.-Russian relations when the U.S. proposed to develop a nuclear defense system unilaterally.

Nuclear Heirs

For U.S. policymakers, one of the must pressing questions was the fate of the vast Soviet nuclear weapons arsenal. Yeltsin presumed to speak for the other Soviet successor states as well as Russia on questions of nuclear weaponry and to arrange for transfer of the nuclear weapons in Ukraine, Belarus, and Kazakhstan to Russia. President Leonid Kravchuk of Ukraine challenged the authority of Yeltsin to negotiate on behalf of the other three

CIS nuclear states, and in mid-March 1992 suspended the transfer of tactical nuclear weapons to Russia, insisting on prior guarantees that the weapons would be destroyed under the supervision of an international commission.[4] In early April 1992, U.S. Secretary of Defense Richard Cheney received assurance from the four nuclear republics of the CIS that all except Russia would be nuclear-free by 1994. But Ukraine opposed acceptance of Russia's right to sign the treaty with the U.S. on behalf of all nuclear successor states until Secretary of State James A. Baker III warned that the U.S. would provide future aid "to those states that demonstrate commitment to freedom, democracy and free markets, as well as nuclear security."[5] The inference was clear: Nuclear security could best be guaranteed by having Russia alone the successor to the Soviet Union as nuclear power, and opponents of that policy would forfeit U.S. aid.

Even as efforts to control nuclear proliferation among the new states of the former Soviet Union were threatened, a new challenge appeared with the news of an agreement between Russia's *Glavkosmos* space agency and the Indian Space Research Organization. Russia had agreed to provide equipment for a rocket capable of launching satellites into near-Earth orbit worth some $400 million. The U.S. charged that the agreement conflicted with the Missile Technology Control Regime (MTCR) and threatened economic sanctions if it was not canceled. The Russian response was in every way conciliatory. While noting that Russia had not been a signatory of the MTCR, and describing the agreement with India as entirely peaceful, it agreed to enter consultations with India, meanwhile suspending the transfer of rocket and space equipment. The Indian case would be the first of several such disputes between Russia and the U.S. over the following decade. Conflicts over issues of arms sales and nuclear proliferation were much complicated by the pressure on the Russian government to support the efforts of its nuclear and defense industries to secure revenue from foreign sales as acquisitions from their own government collapsed.

Despite the conflict over the Indian deal, presidents Yeltsin and Bush produced a significant nuclear arms reduction agreement at their meeting in Washington in June 1992, providing that the total stockpile of Russian and American strategic nuclear arms would be reduced from 21,000 warheads to 6,000–7,000 by 2003. Discussions went beyond reduction of warheads, including the possibility of replacing the ABM Treaty of 1972 with a joint project for defense against nuclear attack. Yeltsin's speech to a joint session of the Congress was interrupted eleven times by standing applause.

The vision of the new relationship was expressed in the "Charter of Russian-American Partnership and Friendship" signed by the American and Russian presidents at their June meeting, which outlined an impressive agenda of Russian-U.S. and "Euro-Atlantic" cooperation in matters of arms and political conflict. Noting "the indivisibility of the security of North America and Europe," it stressed the linkage between the North Atlantic Cooperation Council (NACC), NATO, the West European Union (WEU), and the Conference on Security and Cooperation in Europe (CSCE). The document proposed including the Commonwealth of Independent States in the company of organizations responsible for "maintaining security and peace in this region." The charter also emphasized the importance of cooperation to prevent the proliferation of weapons of mass destruction: The two sides will work toward strengthening and improving regulations governing the nonproliferation of weapons of mass destruction, including nuclear, biological and chemical weapons, missiles and missile technology, as well as destabilizing conventional weapons.[6]

It was agreed that this effort would be pursued both bilaterally and in such multilateral organizations as the Coordinating Committee on Multilateral Export Controls. A major feature of the Washington agreements—proposed by Yeltsin in January and closely linked to his proposals for protection against nuclear proliferation—was the agreement on a global defense system (GDS). Though immensely important and innovative, it was overshadowed by the agreements on reduction of nuclear arms. Writing about this agreement a few weeks later, two distinguished Russian arms specialists noted that

> . . . the proliferation of weapons of mass destruction has become a *fait accompli*. Indeed one can only guess how many countries will possess these weapons within a few years and how prepared they will be to use them without much hesitation.[7]

The authors argued that conventional methods of preventing technology transfer could be effective only with the cooperation of all nations capable of facilitating proliferation of missiles and nuclear weapons, and that this was impossible to achieve. The Washington agreement, if followed by the creation of an effective GDS (as Yeltsin had proposed), in conjunction with the maintenance of powerful Russian and U.S. nuclear arsenals, would be an important new element both in strengthening the nonproliferation regime and in providing a real guarantee that such weapons would not be used by those countries that already had them or were very close to acquiring them.

To their critics, who argued that the whole scheme (which was not unlike an earlier proposal of President Reagan) was naive about the possibilities of U.S.-Russian cooperation, they responded that the U.S. and Russia were already making "a clear transition to partnership—and eventually to allied relations." There were also other potential advantages to both sides from cooperation on the GDS. Russia could offer very advanced research and development in the field, and the Russian defense industry, desperately in need of funding, would benefit from GDS-related projects. They also argued that to facilitate this effort it would be necessary to take an entirely new view of the ABM Treaty:

> The treaty was signed during the Cold War and reflects the realities of that time. In our view, the treaty in its present form has basically served its purpose. We propose that the treaty be amended in certain ways to make possible the elimination of a number of restrictions that are no longer in keeping with the demands of the times—especially with regard to the testing of individual components of a missile defense.[8]

Clearly their proposal, like that of Yeltsin, was based on the assumption that the nations with the technological capacity to develop an effective missile defense no longer needed it to defend against one another, and that by undertaking joint development of an anti-ballistic missile system, they could protect both themselves and the world from the danger of proliferation of weapons of mass destruction, a process they were unlikely to be able to halt.

A decade later, Russian acknowledgment of the possibility of revising the ABM Treaty would seem surprising, but only if one fails to note that it was suggested in the expectation that nuclear missile defense would be a cooperative project. Yeltsin had challenged the U.S. with a comprehensive new concept of bilateral ties—a full partnership—and with a highly original usage of the partnership to achieve global control of weapons of mass destruction through collaboration on missile defense. Issues of nuclear missile defense would continue to be a central issue of the bilateral security agenda, but Yeltsin's daring vision of full U.S.-Russian partnership in this crucial matter soon evaporated in the face of America's determination to build its post–Cold War security system on a more traditional model, beginning with NATO in Europe, and with Russia left out.

NATO and the New European Security System

The issue of NATO expansion has proven to be one of the most difficult problems of Russian-U.S. relations in the post-communist era. The possibility of the eastward expansion of NATO was delivered to Yeltsin at a poignant moment—on the eve of the parliamentary elections in December 1993, only a few months after the armed insurrection in the parliament and at a time when the whole future of his reform program hung in the balance. Both he and Foreign Minister Andrei Kozyrev had repeatedly indicated their preference for a new European security system based on a restructuring of the

Conference on Security and Cooperation in Europe—not on NATO, the organization that symbolized East-West Cold War confrontation. Moreover, the idea of a new regional security organization had already been incorporated in the earlier U.S.-Russian Charter.

The Foreign Intelligence Service (FIS) head, Yevgenii Primakov, presented an extended analysis of the question of NATO expansion—"Prospects for the Expansion of NATO and Russia's Interests"—in November 1993.[9] He argued the need for "transformation of the alliance from a military-political grouping oriented toward repulsing external threats to an instrument for ensuring peace and stability on the basis of the principles of collective security." He expressed serious concern about the lack of clarity of aims and priorities on two aspects of the current process: "changing the alliance's general purpose and the parallel expansion of its political functions and its geographic scope." The hasty expansion of the alliance's geographic scope (moving up to the borders of the former Soviet Union) could draw it into "complicated processes fraught with the possibility of acute struggle in the East European states." Such developments could well delay "NATO's transformation into a universal peacekeeping and stabilizing force" and, by creating "danger for the Russian Federation's interests . . . reduce the chances for finally overcoming the split in the continent and lead to relapses into bloc politics . . . in which NATO's zone of responsibility would . . . [reach] the borders of the Russian Federation."

Kozyrev appeared to be communicating the Russian concerns effectively to NATO in early December 1993 at meetings of the NATO Council and the North Atlantic Cooperation Council (NACC). Russian press reportage viewed the new Partnership for Peace plan as one that would provide for full cooperation of the countries of Eastern Europe and the CIS on specific issues without membership of the bloc. One reporter quoted Kozyrev saying of bloc expansion that "the idea has been buried" and the decision by consensus included "not only NATO and Russia, but also the Central and East European countries."[10] He was confident that the idea of expanding the NATO bloc had been abandoned in favor of various forms of partnership among all the states under the auspices of the NACC, an arrangement endorsed by consensus at the NACC meeting. He had presented to that body his plan for uniting as equal partners all the leading regional organizations in the Euro-Atlantic space within the Conference on Security and Cooperation in Europe, a plan that included NATO, the CIS, the West European Union, the European Union, and other entities. He noted that the NATO leadership regarded the NACC and the partnership program as a sort of waiting room for NATO membership, but indicated that Russia had no intention of joining the queue.

Neither Kozyrev's arguments, nor an appeal from Yeltsin, could halt the NATO action announced just before the December 12, 1993, elections for the newly created State Duma. Both German Defense Minister Volker Ruhe and Secretary of Defense Les Aspin confirmed that NATO would extend its boundaries eastward and accept new members.

Following elections to the Duma—in which Communist candidates and Nationalists did surprisingly well—Russian discussion of NATO expansion became increasingly negative during 1994. Following Lithuania's request for NATO membership on January 5, 1994, Yeltsin spokesman Vyacheslav Kostikov announced that "granting membership to countries located in immediate proximity to Russia's borders . . . could ultimately lead to military and political destabilization."[11] The *Sevodnya* reporter noted, however, that President Clinton favored giving East and Central European countries associate rather than full NATO membership in the Partnership for Peace Program scheduled for review by the NATO Council a few days later. Clinton was quoted as favoring this plan to avoid a "division of Europe," while Lech Walesa called for nothing less than "immediate admission to NATO, with security guarantees."

Russian commentary on NATO expansion into Central and Eastern Europe presented a formidable list of problems that would ensue. These included joint military exercises along Russia's border, a rearmament to fit NATO standards that would foreclose opportunities for Russian arms industries, and, most important of all, a direct challenge to Russian efforts to form a collective security system for the nine CIS member countries. The latter was a central feature of the Russian Foreign Policy Concept developed just one year earlier and had been the basis of the CIS Collective Security Treaty. It was also argued that the introduction of the Partnership plan would replace CIS collective security with a series of bilateral agreements between NATO and individual CIS countries, bypassing Russia.

Interviewed in Naples at the time of the G-7 meeting, Kozyrev spoke of his persistent problems with the Russian parliament, whose red-brown majority portrayed him as a stooge of the West. They are, he noted, "hysterical about political rapprochement between the West and Russia. They understand that they are losing ground." On the other hand, he also found Western leaders deficient in understanding of the plight of the Russian democrats: "there is a real lack of sensitivity [toward Russia] that strikes me as too much self-assurance, even arrogance."[12]

Among Kozyrev's irritations with Western leaders was the criticism of Russian security actions in the CIS, especially in the Caucasus and Central Asia, criticisms that occasionally made veiled references to renascent Russian imperialism and were offered as justification for giving the security guarantee of NATO membership to states of Central and Eastern Europe. His answer to such commentary was to note Russia's "success in avoiding . . . a repeat of the Yugoslav tragedy on the territory of a nuclear superpower . . . [by not taking] an imperial path. . . [or trying] by force to restore the Union state."[13] He made it clear that Russia did not think it necessary to have an external mandate for peacekeeping actions in the CIS, since such actions were based on mutually agreed needs of sovereign states. At the same time, he spoke positively of the latest WEU report on "Relations with Russia," and urged the eventual establishment of a Russia-WEU Consultative Council.

The essence of Kozyrev's argument was a return to the proposals of Yeltsin in early 1992 for a full strategic partnership of the U.S. and Russia. A succinct and compelling statement of that position in an *Izvestia* article in March 1994 was offered just at the time when in both countries there was a growing body of opinion arguing that such a partnership was neither feasible nor desirable. In fact, U.S. policy on NATO membership and the central role of NATO in European security, with Russia excluded, seemed to have foreclosed the possibility altogether.[14]

Kozyrev's argument was a moving plea for a better understanding by Western democracies of the vital connection between the success of Russian democracy and the full incorporation of Russia in the Western security system:

> There is no sensible alternative to partnership, unless, of course, one considers the possibility of missing a historic chance to form a democratic Russian state and transform the unstable post-communist world into a stable and democratic world to be such an alternative. These two goals are directly connected. They are especially dear to Russia's democrats, who have already—more than once—encountered armed (not to mention vehement political) resistance from the opponents of reform. We will continue to struggle to attain these goals, even if we are not heard or correctly understood by our natural friends and allies, the democratic states and governments of the West.

Kozyrev also cautioned the U.S. against impatience with U.S.-Russian differences on policy issues ("partnership does not mean renouncing . . . a policy of defending one's national interests, or, at times, competition and disputes"). And anticipating a line that would be advanced more aggressively by his successor, he warned against the American temptation "to see only one leading power in today's world."

He could not have made it clearer that a new European security system based on NATO was unacceptable to them. That attitude remained essentially unchanged until the Putin era, though even then the issues of NATO expansion by the admission of new members to NATO from Central and Eastern Europe, and Russia's place in the European security system, remained unresolved.

Phase II: The Primakov Era (1996–1999)

A close observer of the Russian scene, Deputy Secretary of State Strobe Talbott, noted that opponents of a Western orientation of Russian policy used NATO enlargement as "proof that the West is bent on humiliating Russia, keeping it weak, plotting its demise."[15] The basis of these fears was the perception that NATO expansion was a hostile military action directed against Russia, since there was little or no understanding of NATO's post-communist political role. Hence NATO policy had contributed much to the declining popularity of pro-Western political figures such as Foreign Minister Kozyrev and was used by his successor to justify the China orientation of his "multipolar" policy.

Efforts were made by NATO leaders to stress NATO's political role in stabilizing post-communist Eastern Europe rather than building an alliance against Russia. Hence the effort to give Russia formal participation in NATO affairs by the introduction of the Founding Act on Mutual Relations, Cooperation, and Security in May 1997, an act that was soon followed by invitations to Poland, Hungary, and the Czech Republic to join NATO. But though the intention of the Founding Act (FA) was to allay Russian fears about NATO intentions, there were soon serious differences about the purposes of the new agreement. It was clear that the FA, without a full-time support apparatus, and operations centered in *ad hoc* committees and monthly meetings of permanent representatives did not provide what the Russians sought. Further negotiations between Foreign Minister Primakov, Secretary of State Madeleine Albright, and NATO Secretary General Javier Solana produced the Permanent Joint Council (PJC), intended to provide regular dialogue between Russia and NATO.

This effort to provide an institutional structure for dialogue between NATO and the Russians was also a disappointment to the Russian side. There was no clear agreement on the agenda and objectives of the PJC, and the organization was given neither permanent headquarters nor secretariat. Most disappointing was the inability of the Russians to participate in NATO deliberations before bloc positions were formed, an arrangement obviously designed to avoid Russian influence during the decision-making process. The PJC came to be regarded as a talking shop while Russian policy makers concentrated on direct approaches to member countries.

The crisis in Kosovo proved a disaster for the progress of Russian-NATO cooperation and the PJC. Though the Kosovo situation was discussed by the PJC in 1998, and it was agreed that Russia would participate in conflict resolution, it was clear by October that the gap between Russia and the NATO powers was unbridgeable. When the NATO ultimatum to Yugoslavia was rejected, and bombing was undertaken in March 1999, the Russians left the PJC in protest. Their objection was not only to the action itself, but to the fact that it was not validated by a UN mandate, thus diminishing the Russian Security Council vote.[16] Both Russian governmental statements and public opinion polls demonstrated a deeply hostile reaction to the NATO action against Yugoslavia.

Phase III: The Putin Era (2000–)

Primakov's leading role in Russian foreign policy ended with his removal as prime minister in May 1999. As Yeltsin explains in his memoir of the period, his relations with Primakov had become increasingly strained over issues of both domestic and foreign policy. Noting that the Kosovo crisis had created much anti-Western sentiment, he described Primakov as being "quite capable of uniting the politicians who dreamed of a new isolationist Russia and a new cold war."[17] He also indicates that even as he replaced Primakov he intended in the long run to have Putin as his prime minister, but was obliged for political reasons to take Stepashin in an interim role.[18] The choice of Putin for prime minister in August 1999 would eventually bring a very different era in U.S.-Russian relations, but the change had only begun to appear shortly before the traumatic events of September 2001. From August 1999, when Putin became prime minister, until September 11, 2001, there was essential continuity in foreign policy in the official pronouncements. Foreign policy still bore the mark of Primakov, indicated as well by continuity in Foreign Minister Igor Ivanov, who was named to the post in August 1998 and retained the position under Putin.

The new "Foreign Policy Concept of the Russian Federation" published in July 2000, four months into the new era and approved by Putin, was a clear illustration of the continuity. It spoke of "an increasing tendency toward the creation of a unipolar world under the economic and military domination of the United States." It further argued "in the pursuit of solutions to the fundamental problems of international security, reliance is being placed on Western institutions and forums with limited membership, and on weakening the role played by the United Nations Security Council." The main negative characteristic of these solutions was that they sought "to downgrade the role of the sovereign state as the fundamental element in international relations, creating the threat of arbitrary intervention in internal affairs."

Russian policy makers were obviously still smarting from the NATO action in Yugoslavia, noting "attempts to introduce such concepts as 'humanitarian intervention' and 'limited sovereignty' into international practice in order to justify unilateral military actions in circumvention of the UN Security Council."[19] The document also announced the Russian intention "to preserve the territorial integrity of the Federal Republic of Yugoslavia and to oppose the dismemberment of that state." Noting the strong anti-Americanism of many of the document's statements, a Russian journalist observed that it preserved "the late-Soviet philippics of the Primakov policy era when Moscow spoiled relations with practically all the Western capitals and international financial institutions." He was clearly worried that the experience and views of that era were now "incorporated in a concept that carries the status of an official directive" and which, as a foreign policy agenda, was "a script for Russia's capitulation."[20] The writing of a new script would not begin seriously until the events of the second half of 2001, beginning with the Putin-Bush meeting at Ljubljana in June 2001.

Bush and Putin at Ljubljana

The central issue for the Russian president was the role and policy of NATO, an organization in which there was still no effective Russian voice. He spoke bluntly, describing NATO in an interview with an American journalist as "a Cold War relic that will only continue to sow the seeds of suspicion in Europe as long as it excludes its onetime arch enemy."[21] The meeting at Brdo castle gave the two presidents an opportunity to get acquainted (with positive personal rapport developing quickly) and to discuss their policy differences over such acute problems as missile defense and NATO expansion. By this time the Russian arguments were familiar—that repudiation of the ABM Treaty to develop a missile defense system would expand competition in nuclear weapons, and that there was no rationale for expansion of NATO since the Russian threat had disappeared. However, the two leaders agreed to establish working groups on strategic stability and export controls. The initiative provided expanded expert assessments and recommendations for further discussions between them at the meeting of the Group of Eight in Genoa in July, at the autumn meeting in Shanghai of the Asia-Pacific Economic Forum, in Washington, and at the Bush ranch in Texas.

In the immediate aftermath of the Ljubljana meeting, both Russian and American presidential aides discouraged the view that the presidents had removed barriers to Russian-U.S. compromise on familiar problems. Putin's comment to American correspondents in Moscow that modifications to the 1972 ABM Treaty were possible, but that without them Russia would be obliged to withdraw from it as well and deploy missiles with multiple warheads, was quickly corrected by his defense minister, Sergei Ivanov, who insisted that modification of the treaty was unacceptable. In the U.S., Colin Powell explained that Bush's suggestion that Russia and America become allies did not mean a military, political, or diplomatic alliance, while National Security Council chair Condoleezza Rice affirmed that the U.S. would build its missile shield with or without Russia.[22]

For his part, Putin chose a meeting in Moscow with the president of Austria, as an occasion to state that if the U.S. installed a missile defense system Russia "will have the legal right to install three, four, five or more warheads on our single-warhead missiles."[23] The reaffirmation of Russian insistence that a U.S. unilateral withdrawal from the constraints of the ABM Treaty would justify Russian abandonment of agreed restrictions on missile armories was categorical. The ensuing discussion in the Russian press displayed a very mixed response—both endorsement of the president's position and charges that the position was unduly rigid and the economic resources for expansion of Russia's nuclear armory were simply unavailable.

On the very eve of the September 11 events, the distinguished political analyst Andrei Piontkovsky offered a cogent critical analysis and proposal on Russian and American policy toward the ABM Treaty and nuclear missile defense.[24] He noted that extensive debates among the Russian political leadership had produced a growing sense of the need for a new treaty to replace the 1972 ABM Treaty, one that would set agreed parameters for missile defense rather than prohibit it entirely. He noted that current U.S. missile defense plans posed no threat to Russian security and argued "fears that these proposals could evolve and that the system could expand . . . are an even greater argument for proposing international legal restrictions on the parameters of a national missile defense system now."

Unfortunately, in Piontkovsky's view, just at the moment when Russia was considering dropping its rigid position on the ABM Treaty, the U.S. leadership seemed to have adopted a "fundamentalist" conception "that rejects all arms control treaties in principle . . . [in order to] secure for itself the broadest possible choice of unilateral strategic decisions." He argued (quite prophetically), "After two or three months . . . the Americans will announce their withdrawal from the 1972 treaty citing the Russians' intractability." He also predicted that the threatened Russian counteractions to such a move would win their cause little sympathy. Anticipating such an outcome, the wisest course for Russia would be to concede the need for ABM Treaty revisions, obliging the American administration to accept new international discussion and agreements on missile defense and other security issues or face rejection of their policy by their European allies and substantial domestic political opposition.

September 11th and After

The terrorist attacks on New York and Washington brought a swift and bold response from President Putin in a television address on September 25, 2001. He offered the cooperation of Russian special services and the provision of information available in Russia on the whereabouts of terrorists and terrorist training facilities. He also offered the use of Russian airspace by aircraft delivering humanitarian supplies to the area of antiterrorist operations, as well as coordination of Russian support with Central Asian states, including the use of Central Asian airfields, and Russian support in international search-and-rescue operations. He indicated Russian support for the Rabbani government and committed to provision of military equipment and supplies. He appointed Defense Minister Sergei Ivanov to coordinate Russian participation in these areas.

Commenting on these developments somewhat later in the authoritative Russian journal *International Affairs*, Ambassador Boris Piadyshev complimented the American side on the effective preparations for the anti-terrorist campaign, the support of NATO, the agreements reached with the CIS Central Asian states, the support of Pakistan against the Taliban, and the speedy preparations for the campaign in Afghanistan. While acknowledging "we (Russia) cannot allow ourselves to be involved beyond a reasonable point," he expressed views that were doubtless widely shared in official circles:

> . . . the latest events confirm the importance of seeing the situation in Chechnya in the context of struggle against international terrorism. Also, that the September events showed that the real threat comes not from ballistic missiles that are going to be countered with a national missile defense system, but from organized international terrorism that is emerging as a formidable force.
>
> Proletarian internationalism and disinterestedness, thank God, are history now. *Realpolitik* posits that both the general policy objective and specific interests—not least economic interests—are a factor to be reckoned with.[25]

The remarkable U.S.-Russian cooperation following the September tragedy inspired hopes that the many conflicts and anti-American policy formulations of the Primakov years would be overcome. For some the mood was euphoric, as in the early Yeltsin period. The views of prominent Russian political analyst Vladimir Frolov were illustrative:

> It can be said that at the ranch in Texas Vladimir Putin closed the book on the foreign policy pursued by Yevgenii Primakov. What the Russian president has been doing in the international arena since September 11 will go down in history as "Putin's foreign policy."[26]

Frolov's case against the earlier Primakov policy was that a state ranking fifteenth in the world in terms of GDP could scarcely hope to be an independent "pole" in the international arena and would inevitably be a junior partner with dynamically developing states such as China and India, despite its nuclear weaponry and Security Council position. He noted that Putin had continued the "Primakov Line" until mid-2001 but turned decidedly westward at that point, seeking to strengthen Russian relations both with Europe and the U.S., recognizing that the West was the main key to both Russia's economic development and its security, and that even the need to limit American "domination" could be better handled by Russian integration into the West:

> Putin's strategic objective is to restore Russia's role as a great power through liberal market reforms and integration into Western political institutions. His tactical goals are to join with the Europeans in forcing the U.S. to 'play by the common rules' and, as he said in an interview with American reporters, to make Washington a "more reliable partner."[27]

Following the September events, and Putin's commitment to support U.S. action in Afghanistan—both Russian support and that of CIS partners in Central Asia—Russian-American relations appeared to have improved remarkably, and further evidence was provided by Putin's visit to the U.S. and extensive meetings with Bush. In a discussion with the international committees of the State Duma and the Federation Council, and international affairs specialists in the government, Putin described the new relationship with the U.S. as "a program of long-term partnership." He spoke of President Bush's endorsement of the proposed arrangement for closer Russian cooperation with NATO by including Russia in the "group of 20" (19 NATO members plus Russia) for adoption and implementation of decisions. He also foresaw expanded economic cooperation, including American recognition of Russia as a market economy and support of Russian entry into the WTO. He acknowledged, "We still disagree on certain points" but affirmed, "a serious and positive transformation of our relations is under way."[28]

Even before Putin's departure for Washington and Texas, however, there were signs in Russia of critical responses to his cooperation with the U.S. in the war against terrorism. One prominent political journalist, noting the "constructive—almost allied relations between Russia and the U.S.," expressed serious concern about America's casual approach to the treaty system controlling nuclear missiles, its criticism of the ABM Treaty, and Bush's expanding list of enemies, warning that "the era of parity has ended not only *de facto*, in view of U.S. nuclear superiority, but also *de jure*, i.e., in terms of treaties and international law." He favored good relations with America, "but we should not lie down and let it walk all over us, even for the sake of fighting international terrorism." He urged forming a bloc with Europe (and joining with China and India) "when America's pretensions to hegemony become too absolute."[29]

A more positive assessment of the new Putin U.S. policy was provided by Sergei Karaganov, chairman of the Presidium of the Council on Foreign and Defense Policy. Karaganov noted that long before the September catastrophe, Putin had cultivated relations with President Bush, and that the range of his efforts to improve and expand Russia's foreign relations had gone beyond the U.S. to include India, China, Japan, South Korea, and the European Union. With respect to the U.S., Karaganov stressed that Putin was seeking to put an end to the prevailing Russian policy of viewing relations "mainly through the prism of two issues: eastward NATO expansion and the Antiballistic Missile Treaty." To deal with the "real problems of the future," both Russia and America would change their outlook:

> ... we haven't yet left the Cold War era. Since the early 1990s, Russia and the U.S. have frozen themselves in theposition of being half enemies, half partners. This indeterminate status casts a shadow on their entire relationship. Putin has proposed that we get out of this in-between situation.[30]

Karaganov firmly supported the Putin initiatives but was sensitive to the obstacles on both the Russian and the American side. Events of the following months would validate his assertions about the rigid definition of the issues in the binational relationship and the survival of Cold War perspectives on both sides.

In Russia the rigidity was most marked in military circles. The expanding cooperation with the Americans was not well received in the General Staff building on Arbat Square, particularly the introduction of American forces into Central Asia. On the ABM issue the position was that it was better for the Americans to withdraw from the ABM Treaty than for Russia to compromise, and Putin's discussion of plans to cut Russia's nuclear missile potential by about 75 percent was firmly rejected.[31] Even more worrying were the attitudes of retired senior military officers expressed in an appeal to Russian government leaders and legislators across the country. The appeal revealed a deep and uncompromising hostility to every major aspect of the Russian transformation since 1991—especially the end of the empire and the socialist economy—and undisguised hatred not only of Gorbachev and Yeltsin but of Putin himself, and, while demanding restoration of military budgets, dismissed all post-communist reduction of military programs and industries as the work of "people serving the interests of the West."[32]

It is not surprising that Putin chose to meet with representatives of the active military leadership shortly before his departure to America to make commitments on such modest improvements in current military support budgets as he could manage. Nor is it surprising that he was reluctant to accept the American proposal for a "paperless reduction" (without signed treaties) of missile armories as proposed by the American side during his visit, and that he stood firm on rejecting the elimination of the ABM Treaty. Only the long-overdue cancellation of the Jackson-Vanik amendment of 1974—dating from Soviet-era restrictions on Jewish emigration—received full agreement.[33]

The issue of the ABM Treaty was settled in Moscow in December 2001. Discussions between U.S. Secretary of State Colin Powell, President Putin, and Foreign Minister Igor Ivanov following Powell's arrival on the 10th simply reaffirmed the opposed positions, the American side claiming that the treaty obstructed necessary testing for development of a needed missile defense system and the Russians continuing to insist that it was a necessary element of strategic stability, though adding that they were prepared to discuss American proposals for revision of the treaty.[34] In the face of the stalemate on the ABM Treaty, discussions were shifted to reduction of strategic weapons.[35]

On December 13, 2001, U.S. Ambassador Alexander Vershbow presented an official notice at the Foreign Ministry that the U.S. would withdraw unilaterally from the ABM Treaty at the end of the six-month period following notification. Both presidents' comments assured their peoples that the action did not diminish the security of either nation. Putin observed that the action was not "unexpected" and that while it was a "mistake," it did not pose a threat to Russian security. He did, however, emphasize that future armament reductions would have to be "codified," that is, backed by formal treaties. Russian military officials noted that the American national missile defense system would continue to be vulnerable for 15 to 20 years.[36] The reaction to the American action was generally moderate, as a prominent journal noted:

The world had been awaiting that announcement [of U.S. withdrawal from the ABM Treaty] with anxiety for several months now, regarding it as all but the declaration of a new Cold War. But there will be no confrontation between Russia and the U.S. The Kremlin doesn't want one.[37]

Predictably, Communist leader Gennadii Zyuganov accused Putin of capitulation and announced that the real danger of the times was not terrorism but American global expansion. But mainline Russian commentary was moderate. Efforts in the Duma to adopt an appeal to the U.S. Congress to block the withdrawal action failed to pass. But the vice-chairman of the State Duma Defense Committee, Aleksei Arbatov, noted that while the withdrawal from the treaty was not intended to insult or strike a blow to Russia, "the decision to do so will have precisely that effect." He also expressed concern about the unilateralism of the U.S. decision, reflecting a "willingness to leave other countries merely to follow in the wake of American policy."[38]

The predominant opinion appeared to be "that nothing tragic has occurred and that Russia's strategic defense system would be able to stand up against it for at least another 10 to 15 years." To those who objected that the Americans had proven ungrateful for the Russian support in Central Asia and Afghanistan, one leading policy analyst responded "the Americans are fighting and defeating the Taliban while we sit things out on the sidelines."[39] Probably the most serious concerns expressed were that the action would weaken the position of Putin vis-à-vis Nationalists and ultra-leftists and that China would likely respond with efforts to build up its strategic nuclear power, a challenge to Russian security.[40]

NATO Again

Discussions of changing the Russian relationship to NATO proceeded more promisingly than the bilateral relationship with the U.S. on missile defense and the ABM Treaty, though these matters also elicited more negative responses from the U.S. than from its NATO allies.

Russia's strong support of the NATO powers engaged in Afghanistan encouraged advocates of a closer Russia-NATO tie. The speech by Lord Robertson, secretary-general of NATO, at Stalingrad in November 2001 compared the joint struggle of Russia and the Western powers against terrorism with the struggle against fascism in World War II, and affirmed that "Russia and NATO are both essential security players in Europe."[41] British Prime Minister Tony Blair proposed replacing the Permanent Joint Council with a new structure to be called the Russia-North Atlantic Council (called "NATO at 20" by some NATO representatives). Russia would be a participant in such areas as terrorism, peacekeeping, and nonproliferation, but would not have the policy veto of a full member.

The main arguments advanced against NATO membership were that Russia did not meet the standards for membership and that full membership would enable it to paralyze the organization's decision-making process.[42] The growing company of supporters of closer Russian ties to NATO included former Secretary of State James A. Baker III, who warned of the danger of alienating defeated adversaries—as with post–World War I Germany.[43] Others had noted that Russia was already a member of the Council of Europe and the Organization for Security and Cooperation in Europe, and that NATO membership would be a logical next step. But at a crucial NATO Council meeting in early December, the majority vote went against giving Russia full voting membership in NATO, though it would be consulted on a wide range of issues.[44]

At the time, convincing arguments were made that Russia could satisfy the formal admission criteria outlined in the NATO Membership Action Plan (MAP). Opposition arguments emphasized the deficiencies of Russian democracy, though with criticism focused not on the fundamentals of the new democratic governmental institutions but on the conflict in Chechnya and recent media disputes. The argument of Zbigniew Brzezinski against the transitional arrangement for full membership, proposed by Tony Blair, stressed that such a scheme would disable the NATO decision-making process.[45] This argument was questionable concerning the operational implications of the Blair proposal, but even more importantly it emphasized only the differences between Russia and NATO rather than the evidence of growing coincidence of interests, and overlooked the clear inadequacy of the existing PJC arrangement from the Russian standpoint.

Developments in U.S.-Russian relations in December 2001–January 2002, following U.S. withdrawal from the ABM Treaty, were not encouraging either on the NATO issue or on the fate of broader hopes of U.S.-Russian partnership. Disappointment was widespread in Russian press commentary, which offered varied assessments of American and Russian blame for what was seen as yet another collapse of hopes (or illusions) of Russian-American partnership. Only a few months later, the news from the NATO meeting in Reykjavik made it clear that the pessimism was premature. The announcement on May 14, 2002, that the U.S. and Russia had agreed to reduce their nuclear arsenals by two-thirds was followed the next day by NATO acceptance of a plan for making Russia a NATO partner, an accord signed in Rome on May 28, 2002. Putin and Bush signed the U.S.-Russian agreement on reduction of nuclear missiles on May 24, 2002, in Moscow. The commitment to a two-thirds reduction of nuclear missiles on both sides by 2012 meant that the aggregate number of such warheads could not exceed 1,700–2,200 for each party. A declaration included in the formal agreement affirmed, "The era in which the United States and Russia saw each other as an enemy or a strategic threat has ended."[46] Other elements of the agreement included cooperation on a Middle East settlement, reaffirmation of joint counter-narcotics efforts, cooperation on controlling the spread of weapons of mass destruction and missiles, and joint work on reduction of weapons-usable fissile material and the development of proliferation-resistant nuclear-reactor and fuel-cycle technologies. Economic cooperation continued to be a central aim of President Putin, who secured a commitment of support for Russia's entry into the WTO and President Bush's commitment to an American designation of Russia as a market economy to facilitate expanded trade relations, which was formalized in June 2002.

The May 2002 agreement replaced the Permanent Joint Council with the Russia-NATO Council, a body consisting of the nineteen Western NATO members and Russia that would meet to set policy and approve programs. Any of the members of the Council would have the right to veto further action or debate, but NATO members would have the right to consider rejected items separately.[47] It was not clear what would be the agreed-upon scope of the Council's concerns, but obvious ones were counterterrorism policy, ballistic missile defense, crisis management and peacekeeping, arms proliferation, search-and-rescue operations and emergency planning, and joint management of airspace. The scope of Russia's role was uncertain, since the conditions for ending debate at the NATO-Russia Council and returning items to the North Atlantic Council were not clear. Recognizing these limits, President Putin described the arrangement as "only a beginning," though "a major step which would previously have been unthinkable."

Conclusion

By June of 2002, it looked very much as if Yeltsin's vision of Russia as a partner of democratic America, presented to the U.S. Congress a decade earlier, might at last be realized. Looking back over the turbulent years that had passed since the overturn of Communism and the Soviet Union brings to mind occasions when it appeared frighteningly as if the new Russia were failing in its bold effort to build a democracy and market economy amidst the wreckage of the Soviet Union and might again adopt a hostile posture toward the U.S. In such times, the ominous label "Weimar Russia" was used, implying that Russia's democratic effort would be destroyed as Germany's Weimar Republic was by Hitler.

But Russia's revolution, for all its trials over a turbulent and dangerous decade, has peacefully dismantled a vast empire, built a democracy and a market economy, and moved toward firm integration into the economic and security system of the Western democracies. Differences on policy issues between the U.S. and Russia would doubtless recur, as with U.S. NATO allies over many years, but as negotiable differences between partners committed to common principles of government and economy within a cooperative security system.

Notes

1. *Rossiiskaia gazeta*, January 30, 1992, pp. 1–2.

2. *Rossiiskaia gazeta*, February 3, 1992, pp. 1, 3.

3. Ibid.

4. Kravchuk's protest concerning Yeltsin's failure to consult with the other CIS nuclear states was reported under the title "Yeltsin Didn't Consult Me" in *Izvestia*, February 3, 1992, p. 3.

5. *Izvestia*, April 17, 1992, p. 6.

6. *Rossiiskaia gazeta*, June 19, 1992, pp. 1–2.

7. *Krasnaia zvezda*, August 7, 1992, p. 2. The specialists were Sergei Blagovolin, president of the Institute for National Security and Strategic Studies (INSSS), and Ilya Surkov, INSSS vice-president.

8. Ibid. The subject of the level of development of Russian research on missile defense technology has received little attention in the West, though in fact it was highly developed, and probably ahead of U.S. development of related technology.

9. A condensed version of the document was published in *Nezavisimaia gazeta*, November 26, 1993, pp. 1, 6.

10. *Nezavisimaia gazeta*, December 8, 1994, p. 4.

11. *Segodnia*, January 6, 1994, p. 1.

12. *Time Magazine Archive*, vol. 144, no. 2 (July 11, 1994), p. 34.

13. Statement in Paris before the December 1, 1994, Plenary Session of the West European Union, quoted in *Segodnia*, December 2, 1994, p. 1.

14. The article was titled "Russia and the U.S.: Partnership Is Not Premature, It Is Overdue," *Izvestia*, March 11, 1994, p. 3.

15. Quoted by John Feffer in "U.S.-Russian Relations: Avoiding a Cold Peace," *Foreign Policy*, no. 104 (Fall 1996), p. 21.

16. The failure of the PJC is presented in Peter Trenin-Straussov, "The NATO-Russia Permanent Joint Council in 1997–99. Anatomy of a Failure," Berlin Information Center for Transatlantic Security, July 1999.

17. Boris Yeltsin, *Midnight Diaries* (New York: Public Affairs, 2000), pp. 271–272.

18. Ibid., p. 275.

19. "The Foreign Policy Concept of the Russian Federation," *Nezavisimaia gazeta*, July 11, 2000, pp. 1, 6; in *Current Digest of the Post-Soviet Press*, vol. 52, no. 28 (2000), pp. 6–8 [hereafter *CDPSP*].

20. *Novye Izvestiia*, July 11, 2000, p. 3; in *CDPSP*, ibid.

21. Peter Baker, "Putin Offers West Reassurances and Ideas on NATO," *Washington Post*, July 18, 2001, p. A-1.

22. *Nezavisimaia gazeta*, June 22, 2001, p. 1; in *CDPSP*, vol. 53, no. 25 (2001), p. 4.

23. *Vremia novostei*, June 25, 2001, p. 1; in *CDPSP*, ibid.

24. *Nezavisimaia gazeta*, September 4, 2001, p. 6; in *CDPSP*, vol. 53, no. 36 (2001), p. 5.

25. Boris Piadyshev, "After the Terrorist Attack in the U.S.," *International Affairs*, vol. 47, no. 5 (2001), pp. 6–7.

26. *Vremia MN*, November 23, 2001, p. 5; in *CDPSP*, vol. 53, no. 47 (2001), p. 5.

27. Ibid.

28. *Trud*, November 24, 2000, p. 4; in *CDPSP*, vol. 53, no. 47 (2001), p. 8.

29. *Vremia MN*, November 17, 2001, p. 3; in *CDPSP*, vol. 53, no. 46 (2001), p. 12.

30. *Obshchaia gazeta*, November 22–28, 2001; in *CDPSP*, vol. 53, no. 46 (2001), p. 6.

31. *Nezavisimaia gazeta*, November 13, 2001, pp. 1–2; in *CDPSP*, vol. 53, no. 46 (2001), pp. 8–9.

32. *Sovetskaia Rossiia*, November 10, 2001, p. 3; in *CDPSP*, vol. 53, no. 46 (2001), p. 4.

33. *Vremia novostei*, November 15, 2001, pp. 1–2; in *CDPSP*, vol. 53, no. 46 (2001), p. 1.

34. *Rossiiskaia gazeta*, December 8, 2001, p. 7; in *CDPSP*, vol. 53, no. 49 (2000), p. 20.

35. *Kommersant*, December 11, 2001, p. 1; in *CDPSP*, vol. 53, no. 50 (2001), pp. 1–2.

36. *Izvestia*, December 14, 2001, p. 1; in *CDPSP*, ibid.

37. *Kommersant*, December 14, 2001, p. 1; in CDPSP, ibid., p. 3.

38. *Nezavisimaia gazeta*, December 14, 2001, pp. 1–2; in CDPSP, vol. 53, no. 50 (2001), p. 8.

39. *Nezavisimaia gazeta*, December 15, 2001, pp. 1, 3; in *CDPSP*, vol. 53, no. 50 (2001), p. 7.

40. *Kommersant*, December 15, 2001, p. 8; in *CDPSP*, vol. 53, no. 50 (2001), p. 7.

41. Speech titled "NATO and Russia: A Special Relationship" given at Volgograd Technical University, November 22, 2001.

42. See, for example, Ronald Asmus and Jeremy Rosner, "Don't Give Russia a Veto at NATO," *Washington Times*, December 5, 2001, p. 5.

43. "Russia in NATO?," *The Washington Quarterly*, vol. 25, no. 1 (Winter 2002), p. 12.

44. *Rossiiskaia gazeta*, December 8, 2001, p. 7; in *CDPSP*, vol. 53, no. 49 (2001), p. 20.

45. Zbigniew Brzezinski, "NATO Should Remain Wary of Russia," *The Wall Street Journal*, November 28, 2001, p. 4.

46. *The New York Times*, May 25, 2002, p. A7.

47. *The New York Times*, April 23, 2002, p. A3; ibid., May 29, 2002, pp. A1, 7.

For Further Reading

Leon Aron and Kenneth M. Jensen, *The Emergence of Russian Foreign Policy* (Washington, DC: The United States Institute of Peace, 1994).

Leszek Buszynski, *Russian Foreign Policy after the Cold War* (Westport, CT: Praeger Publishers, 1996).

Nicolai N. Petro and Alvin Z. Rubinstein, *Russian Foreign Policy: From Empire to Nation State* (New York: Longman, 1997).

Robert H. Donaldson and Joseph L. Nogee, *The Foreign Policy of Russia* (Armonk, NY: M. E. Sharpe, 1998).

Leo Cooper, *Russia and the World* (New York: St. Martin's Press, Inc., 1999).

Igor S. Ivanov, *The New Russian Diplomacy* (Washington, DC: Nixon Center and Brookings Institution Press, 2002).

PART TWO

SOCIOPOLITICAL
CHALLENGES

6

The Challenge of
Crime and Corruption

Louise I. Shelley

Many writings on Russian organized crime refer to it as a single phenomenon. Although organized crime in Russia and the former Soviet states share many common attributes, there are significant regional variations. In contrast with the Sicilian Mafia, which developed on an island, Russian organized crime developed over the vast territory of the former Soviet Union.[1] Therefore, it is not as uniform a phenomenon as the Mafia or as centrally controlled. The highly differentiated phenomenon of contemporary Russian organized crime is explained by the vast territory of Russia, the presence of crime groups from all regions of the former Soviet Union operating within Russia, and the increasing visibility of international crime groups exploiting the absence of effective law enforcement.

In the mid-1990s, in an article entitled "The Criminal Kaleidoscope," this author suggested that post-Soviet organized crime was evolving differently in the various regions of the former USSR. This evolution is in large part explained by the Soviet legacy combined with the impact of organized crime groups' trading partners.[2] At the beginning of the twenty-first century, this observation has become more apparent than it was at that time. Those engaged in the illicit trade are influenced by their partners in crime, just as legitimate businesses are influenced by their legitimate partners. This insight becomes more apparent the farther one moves from Moscow and the control of the central state.

Without an effective law enforcement apparatus and with little civil society to curb the rise of organized crime, the criminal trajectories set in motion in the early post-Soviet period have continued without impediment. Organized crime continues to drain money from the economy while reinvesting only a small share of its proceeds; it achieves its objectives through massive collusion with and corruption of politicians.[3] Moreover, organized crime has a raider mentality toward Russia's resources–both natural and human. Yet the resources that are exploited differ considerably by region of the country, and consequently the demands of organized crime's trading partners also vary.

Methodology

The analysis of this paper is based on a wide variety of sources, relying heavily on the criminological research produced at TraCCC's (the Transnational Crime and Corruption Center) affiliated centers located in Moscow, St. Petersburg, Yekaterinburg, Irkutsk, and Vladivostok. These centers undertake collaborative research and sponsor grants to scholars in their regions to analyze major problems of organized crime.[4] Regional press analyses evaluate regional trends. The multidisciplinary research also focuses on in-depth analysis of particular categories of organized crime. National data of the Ministry of Interior has been used to assess regional differences in organized crime.

In addition, econometric analysis has been used to analyze the rates of violent crime in the country and to correlate these results with economic and political indicators. These data reveal striking differences among regions, with crime rates rising from west to east.[5]

Overall Crime Trends

Despite significant regional differences in crime, there are certain important common trends over the last decade that span Russia's diverse regions. Some of these are consequences of the Soviet period and others are a direct result of specific policies undertaken during the period of economic transition.

1) Crime rates grew in the 1990s, a trend that began very strongly at the end of the Soviet period;
2) Crime statistics do not accurately reflect some of the crime trends, as they minimize governmental corruption and the theft of government property;
3) Violent crime rates rose dramatically, a consequence of the transition and the rise of organized crime;
4) Youth crime rose dramatically in the first years of the transition; and
5) The criminal justice system targeted only low-level offenders and individuals from the lowest social class in Russia.

Crime grew in the final years of the Soviet period, but increased markedly in the initial years of the transition from Communism (1990–1999). During these years, the average annual growth of registered crime increased by more than 7 percent, whereas the population not only did not increase but fell by an average of 0.15 percent per year. In 1999, crime increased by 16 percent, and the population fell by 0.3 percent. The most notable increase in crime was in the crimes linked to the new market economy, where the growth rate was more than double that for other categories of criminality.[6]

Certain characteristics of the crime rates for this period deserve particular attention. For example, in the transitional period of the 1990s, reported rates of theft of personal property exceeded those for state property. The contrary might have been expected, because during these years theft from enterprises through embezzlement and asset-stripping reached its height. The rise of organized crime did contribute to an increase in burglaries and other property-related offenses. Citizens reflected their heightened insecurity by reporting these pervasive property crimes to the police. Yet the massive theft of resources from state institutions, banks, and enterprises remained unreported because of the absence of economic crime enforcement, explained in part by the complicity of law enforcement with the perpetrators of these crimes.

Youth crime rose dramatically in the first years of the 1990s but declined in the second half of the decade.[7] Some of this growth may be ascribed to the rise of homelessness among youth and the increasing problems of alcoholism, drug use, unemployment, and domestic violence which accompanied the difficult initial years of the transition. While these problems did not go away in the second half of the 1990s, the intense growth in youth crime seen in the first post-transition years did not continue. Adult crime rates, however, did not level off in the second half of the 1990s. Instead, crime rates for violent acts committed by adults grew dramatically both in the domestic and the public sphere. The killings associated with organized crime and banditry were notable. Certain large cities with significant crime groups such as Moscow, St. Petersburg, and Ekaterinburg saw many killings associated with the division of territory among crime groups.[8]

Violent crime not associated with organized crime activity also grew notably. Part of this is explained by the availability of weapons, which were tightly controlled during the Soviet period.[9] The availability of arms, facilitated by the small-weapons trade of Russian organized crime and former military personnel, made many ordinary acts of crime more violent than in the past.[10] The decline in Russian medical care meant that many individuals who might have been merely assault victims in the past were now homicide victims. Consequently, Russia had 27.4 mortalities from homicides and assaults per 100,000—a rate that is four times that of the United States and approximately twenty times that of Western Europe.[11]

The post-Soviet criminal justice system prosecuted few offenders from the criminal or political elite. Instead, most prosecuted individuals came from the lowest economic and social level of society; few were members of significant criminal organizations. Most serious offenders enjoyed immunity from arrest. Some enjoyed this privilege through their election to national and regional parliaments. For others, their immunity stemmed more from their manipulation of the criminal justice system than from the institutionalized protections granted to wealthy politicians. Criminals with significant assets or allied with affluent crime groups could bribe themselves out of criminal arrests and investigations. Lawyers could also help find technicalities leading to the release of suspected offenders.

Court data and that on the incarcerated population reveal that the criminal courts and labor camps were reserved for the lowest strata of Russian society. Of the 1.7 million offenders charged with crimes in 1999, 56 percent were persons with no steady income, 27 percent committed a crime while under the influence of alcohol, narcotics, or other substance, 15 percent were committed by women, and 11 percent were committed by minors. These are the most defenseless and poorest members of society. The highly significant detention rate of women is further evidence of the impact of the impoverishment of women in the transitional period.[12]

Overall Organized Crime Trends

The growth in crime and the decline in the effectiveness of law enforcement have been widespread in Russia. Recalling the enlightenment-era figure Cesare Beccaria's comment that "the certainty of punishment was more important than its severity," the absence of certainty had a most pronounced effect in certain areas of crime. Moreover, certain important trends in organized crime development had a marked impact on the crime rate. These important trends are as follows:

1) Impunity has a particularly strong impact on certain categories of crime;
2) Contract killings are still common;
3) Organized-crime investment in the legitimate economy has not transformed them into legitimate investors;
4) A culture that permits the perpetuation of crime;
5) Great diversification of organized crime activity in all regions of the country;
6) Strong political involvement of organized crime in politics;
7) Capital markets are heavily dominated by organized crime;

8) Very wide variety of money laundering techniques to support and sustain organized crime activity;

9) Presence of organized crime groups from many different parts of the former Soviet Union operate in many different regions of Russia.

Organized crime was able to grow so rapidly in Russia in the first decade of the post-Soviet period because of the pervasive corruption of government officials and their recognition that they could act with almost total impunity. During the Soviet period, party sanctions placed some curbs on governmental misconduct; but with the collapse of the Communist Party, and in the absence of respect for the rule of law, there were no limits on the conduct of government officials. The crime groups could expand their scope of operations enormously because they had cooperation from government officials and faced no sanctions from the law enforcement apparatus. Corruption, bribery, and abuse of power escalated rapidly, but there was a sharp diminution of prosecutions for these offenses.[13]

The law enforcement apparatus decimated by poor morale and dangerous work conditions, as well as the dismissal and departure of many long-term personnel, was ill-equipped to deal with an increasing number of serious crimes. Their inexperience with investigating and prosecuting the economic crimes of a market economy gave organized crime groups the opportunity to expand their financial reach enormously. The inability of law enforcement to solve the numerous contract killings meant that there has been no state curb on the violence of organized crime. There has been a decline of mass shootouts among criminals, but many contract killings remain unsolved throughout the country. This change in patterns of violence is not a consequence of state intervention. Instead, it represents an evolution in the criminality of the crime groups. Once the initial division of state property was completed and territory was divided among crime groups, there was not mass warfare among rival gangs. The violence of organized crime is now directed more at competitors, and strikes at leading business people of important sectors of the economy such as the ports in Vladivostok and markets in Moscow. Violence against investigative journalists continues as a means of curbing investigative reporting.[14]

Organized crime made significant investments in the legitimate economy during the initial stages of privatization in the early 1990s.[15] They have diversified these investments throughout the decade; but despite the large stakes they now hold, they are not behaving as legitimate businessmen in many areas of their business operations. They still use force and threats of force to enforce contracts and to force employees to sell their shares to the crime group at reduced prices.

The diversity of post-Soviet organized crime is one of its hallmarks. Organized crime groups are not involved exclusively in one area of criminal activity. Certain crime groups may specialize in drug trafficking, arms trafficking or auto theft, but most crime groups are multifaceted, spanning many aspects of the legitimate and illegitimate sectors of the economy simultaneously. In any one region of the country, all forms of illicit activity will be present. The illicit sectors might be divided among different crime groups, although turf wars are still fought occasionally over territory and control over various sectors of the economy.

Crime groups are able to have such economic power because they have influence over power structures and in many regions are fully integrated into the existing power structures. Organized crime groups are involved in politics at the local, regional, and national levels. Crime groups fund candidates for office, run themselves for office, and in the most criminogenic regions have their own representatives at a very high level in the mayor's office and at the regional level. This has allowed organized crime to impede the adoption of needed legislation and in some cases to shape the laws to its own ends.[16]

The commission on pardons was dissolved by President Putin in 2001 and its functions were handed over to regional administrative bodies. While some thought there were political motives for the elimination of the commission, subsequent investigations revealed massive corruption occurring in this body. A pardon cost $5,000 for each year of a reduced sentence, and many very serious pardoned recidivists subsequently committed very serious offenses. It is not clear whether the criminals made payments directly to members of the commission or through intermediaries. Many murderers, bandits, and kidnappers were pardoned after such payments.[17]

Particularly pernicious has been organized crime's control over financial markets. There are few analogous situations to this in other countries. The influence that Japanese crime groups have exercised over the banking sector has been devastating, but their influence is less than that of comparable Russian groups. The control that they have exercised has contributed to the collapse of the financial sector, the most notable of which was in August 1998. But on a daily level, this control has impeded the development of legitimate businesses. There have not been sources from which honest entrepreneurs could borrow to obtain legitimate capital. Therefore, they have been dependent on organized crime, which has lent businesses funds at usurious rates and used this as leverage to take over viable businesses.

Citizens have lost their life savings in the organized-crime-infiltrated banking system by investing in joint stock companies, only to have their assets diluted through schemes perpetrated by so-called "investment funds." This has led to the impoverishment of many in society and has made citizens lose any faith they might have had in the capitalist system. Moreover, because of the loss of their capital, citizens have been deprived of the assets they needed to start and maintain legitimate businesses.

The control that organized crime has had over the banking sector and exchange operations has facilitated massive money laundering out of and through Russia. Russian specialists analyzing money laundering in their country have identified dozens of techniques organized crime has used to exploit its control over the import-export sector, rare and precious metals markets, casinos, and even transport to move funds. Russian money laundering, as distinct from just capital flight, has been so significant that it has drained Russia of much of its investment capital.[18]

The ease with which foreign organized crime groups can launder their money in Russia is just one reason that they have become such significant actors on Russian territory. Many groups from areas of the former Soviet Union such as the Caucasus operate in Russia. Apart from these groups of the "near abroad," many Asian crime groups are now operating in the Russian Far East. The massive corruption in the Far East government and the incapacitation of law enforcement has provided the opportunity for many groups to operate with impunity. Such diverse groups as North and South Koreans, Chinese, Japanese, and Vietnamese operate in this region, as will subsequently be discussed more fully. Nigerians recruit drug runners in St. Petersburg, Italian organized-crime groups launder their money in Ekaterinburg, and Latin American groups distribute drugs to and through various regions of the former Soviet Union.

Russian organized-crime groups are themselves multiethnic and represent cooperation among groups that are often antagonistic outside of the criminal world. Foreign crime groups not only operate on Russian territory but often form partnerships with Russian crime groups to carry out their crimes in particular areas.

In the past decade, the crime picture in Russia has become vastly more complicated than it was during the Soviet period. Russia has become integrated into the illicit global economy in many different ways. The regional and national differentiation of the crime has made Russian organized crime among the most diversified in the world today.

The Diversification of Russian Crime

Significant regional differences exist in crime. In the 1970s, Soviet scholars were already noting that crime rates in the new cities of Siberia and the Urals had higher crime rates than those favored urban centers that could control their population influx. This geographic difference is even more pronounced

Figure 6.1 **Regional Distribution of Crime 2001**

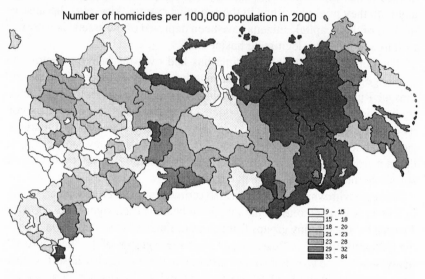

Number of homicides per 100,000 population in 2000

☐	9 – 15
☐	15 – 18
☐	18 – 20
▨	21 – 23
▨	23 – 28
▨	29 – 32
■	33 – 84

Source: Y.V. Andrienko, constructed from data from *Regioni Rossii: Statisticheskikh sbornikh v 2 tomakh* (Moscow: Goskomstat Rossii, 2001), p. 827.

today. Crime rates escalate consistently as one moves from the western part of the country east. The explanation for this is not that colder parts of the country have higher crime rates. Instead, this phenomenon is a legacy of Soviet-era policies of strict population controls, a massive institutionalized criminal population, and a policy of central economic planning that failed to consider the needs of citizens.

In many parts of the world, there is a strong correlation between the level of urbanization and crime rates. The policy of exiling recidivists to rural areas meant that crime rates in rural areas were often high. In 2001, rates for homicide, assault, rape, and hooliganism were higher in more rural areas but for "open stealing" (stealing without weapons) and armed robbery were lower.[19] It is these latter crimes that are particularly associated with the urban environment because of the greater presence of valuable property to steal in cities, whereas violent crime is still characteristic of rural areas.

During the Soviet era, new industrialized cities were established east of the Urals, particularly in Siberia. These new cities were populated primarily by young men, and there was no planning for employment that would attract women to these same communities. No infrastructure was established to provide recreational activities for the new inhabitants. Without a private sector, there was no group that could offer alternatives or substitute for the absence of state services. With the existing internal passport system and rules of mobility, women could not move to these communities without employment. Therefore, these new cities quickly became areas with high rates of alcohol consumption, violent crime, and other forms of criminality.

Compounding the crime rates in these areas were other problems associated with the *propiska* (registration) and passport system and rules determining personal mobility. Under this system, young males could not register in major cities such as Moscow and Leningrad, as well as many others. In unpublished regulations enforced by the police, individuals who were sent to labor camps for a large variety of offenses lost the right to return to their home communities. Moreover, they were prohibited from settling in almost any significant-size cities in the former Soviet Union. With these restrictions on their post–labor camp release, many ex-offenders settled near the labor camps from which they were released.[20] Therefore, the Urals and Siberia and parts of the Russian Far East absorbed very large numbers of hardened criminals who continued to commit very serious crimes, often in association with their fellow ex-convicts. They continue to recruit younger males into their criminal subculture.

The historical Soviet legacy of high concentrations of serious offenders in the Urals and Siberia contributes to the continued rate of serious crime and organized criminality in these regions. Russia has been different from most countries in that the level of criminality is not positively correlated with the level of urbanization. In many cases, it is the secondary and tertiary cities with the highest crime rates or even rural areas in which hardened criminals have been forced to settle that have the highest crime rates.

The ethnic conflict in the Caucasus has affected crime rates and patterns, particularly in the south of the country. Arms and drug trafficking and kidnapping are particularly pronounced in this region, and local business people are pressured to launder money for the Chechen crime groups. The instability in Georgia is exacerbating the criminal situation in the region, as criminals can find safe haven there.

Siberia and the Urals

Siberia and the Urals have crime problems that bear many similarities but are distinctive from other regions of the country. Both have highly criminogenic populations, a legacy of the Soviet-era policies previously described. Rich natural resources in these regions are sources of high profits for crime groups. Crime groups in both areas have made enormous profits exporting precious metals, gold, and diamonds from Russia. Furthermore, the Urals region was a major center of the Soviet Union's military-industrial complex. The enormous wealth generated from these enterprises was a natural target for organized crime groups in the transitional period.[21]

The Siberian region, in contrast to the central and European parts of Russia, has many fewer registered organized crime groups. There are about 5,000 active participants in organized crime in the western and eastern Siberian regions, an area that represents about one-third of Russia's territory. In the Siberian region, the rate of growth of organized crime groups was significant. The growth in the Siberian region parallels that which is occurring in Russia as whole.[22] Between 1994 and 1999, the crimes attributed to organized crime groups rose 1.7 times. This growth is a consequence of two factors—the growing professionalism of the criminals and the declining effectiveness of law enforcement.

Part of the impotence of law enforcement is explained by the close links that exist between crime groups, local officials, and police and judges. Twelve mayors and deputy mayors and sixty-seven law enforcement officials were placed on trial for abuse of their positions, significant numbers in a period when there has been almost total immunity for government officials.[23]

In the absence of effective state enforcement, private enforcement has emerged on a massive scale. Criminal protectors compete with state law enforcement and legitimate private security firms. But some of this private enforcement draws heavily on the corrupt and criminalized structures of the control apparatus for its personnel. Members of private security firms in the Urals have murdered individuals of rival criminal gangs who have failed to pay their debts.[24]

A distinctive feature is the very serious crimes committed by organized criminals. There has been enormous growth in the most serious crimes, and they make up a larger proportion of overall crime commission than is generally observed. The most widespread of their criminality is in the arms trades, but they are also very involved in kidnapping and banditry.[25] These crime groups are still close to the patterns of behavior of the labor camps with which they are closely associated. The presence of arms and banditry gives a visibly violent character to the crime that is heavily reported in the mass media.[26] This is crime less tied to the legitimate economy than in the Urals region.

Less reflected in the crime statistics are the very serious and pervasive crimes tied to organized crime groups. The absence of effective enforcement makes it very difficult to address the rise of auto thefts and drug trafficking. Some of this crime is connected with crime groups in other regions in Russia and even with foreign crime groups. The region around Irkutsk is at the epicenter of a drug epidemic as reflected in the high number of drug addicts and HIV cases, a problem to which local authorities are paying insufficient attention.[27]

In the Urals, crime groups have assumed a different formation. At the end of the Soviet era, Ekaterinburg was the third largest city in Russia and a center of Communist Party power. President Yeltsin came from Ekaterinburg as did many top party and legal officials. Many of the officials of the Yeltsin era have survived to be officials in Russia. In the Urals, there has been a particularly strong melding of the Communist-era elite with the elite of the criminal population. Representatives of the leading criminal organization in the region, Uralmash, are present in the governmental structures even at the regional level.[28]

The banditry and large number of contract killings that characterized Ekaterinburg and nearby regions was an important feature of the criminal environment in the early 1990s. Now, however, the crime patterns have changed. Many of the serious offenders of the early 1990s have died, and a significant share has turned to the most serious forms of traditional criminal activity such as drugs, arms trafficking, prostitution, and gambling. While a significant share have entered into the less violent side of highly profitable organized crime—import-export business, money laundering, and exploiting the valued properties they acquired through privatization in the early 1990s. This movement into the legitimate economy does not mean that the criminals have become legitimate.[29] Violence or threats of violence are still used to enforce contracts, intimidate shareholders, and secure and enforce monopolies.

In one factory visited by the author in a region in the Urals under the domination of a very strong international crime group, the employees were deprived of their shares after the collapse of the ruble in 1998. The crime group allowed limited employee ownership until the late 1990s. After the ruble's collapse, when Russians could no longer afford to import food and liquor on a mass scale, domestic production escalated in value. The crime group that had allowed employee ownership when the plant was of little value took a different stance after 1998. Workers were then forced to sell their shares at a sharply reduced price to the criminal organization.

The particular patterns of Russian privatization have strengthened the links between crime groups and the ruling elite. Under one favored form of privatization, 51 percent of the shares of state enterprises were retained by the state and 49 percent of the shares were available to the public. Consequently, in a region like the Urals, the governing board of a factory or enterprise could consist of the local government and the crime group. The association and integration of the government and the crime group in the legitimate sector of the economy means that there are frequent and consistent interactions between government and the crime groups, which can then continue in other areas outside of the individual enterprise where these relations are regularized. Moreover, the relations that exist on the governmental level facilitate the contacts in the business sphere.

The difference between the state and the crime groups is harder to distinguish in parts of the Urals. Individuals with established criminal pasts are entering politics on both the local and national levels.[30] The integration of the criminal elite and the political elite may be greatest in this region and the Far East, which will be discussed subsequently.

The criminality of this region, despite its inland location, has managed to internationalize on a global scale. Without ports and with access only to railroads and air transport, the crime groups have used telecommunications extensively to support their activities. Aiding the internationalization of this crime is the fact that many ethnic Germans from the region have been able to resettle in West Germany. The links that exist between the Urals and Western Europe have facilitated a massive import-export trade— billions of dollars of it being illicit. The most common method used to facilitate this illicit export of valuable commodities is the registration of export contracts with the authorities in the region. Export documents provide a three-year period for payment. The bogus firms established overseas by the Urals crime groups that receive these precious commodities in their European shell companies never pay for the goods, and the debt is written off after a three-year period.

The Russian Far East

The Russian Far East has become a central locale for the operation of organized crime and terrorists from North Korea because it is far from the central controls of Moscow, and the governmental and law enforcement structures are heavily infiltrated by organized crime.[31] The enormous difference in the standard of living between the citizens of the Russian Far East where monthly income averages less than a $100 a month and neighboring South Korea and Japan with high income levels provides a fertile ground for illicit activity of both the criminal and political kind. This outcome is particularly true when one considers the weak political controls and the criminalized elite that operates in many regions of the Russian Far East. Putin's government has had little success or even motivation to change the power structure in the region.

Organized crime groups from the Russian Far East work with North and South Koreans, Japanese, Chinese, and Vietnamese, among others, who can operate with impunity on Russian territory in the region around Vladivostok. Much of the criminality is connected with the ports and the massive shipping that emanates from this region. Many of the shipping companies and fishing companies are controlled by organized crime. Because their primary purpose is not to ship legitimate commodities, they can afford to undercut the rates of legitimate shipping companies operating in the Pacific Rim. Some of the ships are used to transport drugs, arms, and other illicit goods. Joint ventures with American companies have suffered significantly because the partner was organized crime. In one case in the mid-1990s, an American sailor on the ship of one of the joint ventures was killed after discovering the criminal nature of the Russian partners.

The international community in recent years has become more aware of the crime-ridden maritime industry of the Russian Far East. The killing in late 2001 of Leonid Bochkov, the general director of the Eastern port in Vladivostok, is only further evidence of the continued problem. His death was seen as part of a conflict for spheres of influence by different crime groups operating in the region.[32]

The impoverished military has contributed to massive unauthorized arms sales to foreign governments and organized crime groups. Top-level classified military technology was sold to the South Korean military as "scrap metal," and the perpetrators of this crime placed on trial in the Vladivostok courts. A sale of Russian helicopters to North Koreans was averted only when members of the police, who were not part of the scheme, stumbled on the helicopters just prior to delivery.[33]

Much of the crime is connected with exploration of natural resources. Japanese crime groups contract with Russian crime groups and corrupt local officials to obtain timber for Japanese markets. Endangered species are sought by Asian markets because of their believed value as aphrodisiacs. Particularly damaging to the economic health of the region is the theft of the valuable fish harvest by organized crime and corrupt officials. It is estimated that 90 percent of the revenues from the fishing in the region are diverted, and the loss is estimated at $350 million annually.[34]

The valuable oil resources of the Russian Far East remain untapped because of the endemic crime and corruption. Foreign investors, in particular those from Alaska and Western Canada, with experience working in the oil industry in Northern regions, have withdrawn. Faced with massive corruption, intimidation, and criminal partners, many have abandoned their initial investments in the region.

The Russian Far East has been the beneficiary of many valuable financial transfers from the central government to the region. The theft of state budgetary resources has left the region without the resources to maintain its infrastructure. Highly symptomatic of this has been the freezing of pipes that deliver heat and water to many communities in the Far East. Electricity is rationed and delivered only hourly daily. In the terribly difficult conditions of cold without heat, there has been a massive exodus from the region. Within the past decade, over 10 percent of the population has departed, and in many Far East communities there is massive unemployment.

The unemployment has particularly affected women, who have been very vulnerable to the human traffickers who operate in the region. There has been a massive rise in prostitution. In Khabarovsk, the number of brothels has grown to over one thousand. Moreover, there has been a significant rise in trafficking of Russian women from this region to Korea, Japan, and China. The first two are less frequent because of the higher costs of transport and also the costs of obtaining entry documents for the women. Among those working under registered businesses, the figures are different. According to the internal affairs department of the Khabarovsk region last year, there were twenty-seven firms specializing in finding foreign employment, primarily in Japan and Korea, in show business.[35] Many more women went to China under different terms, because border crossings are easier between Russia and China than other Asian countries.

Moscow and St. Petersburg

Moscow as the economic powerhouse of Russia has the full range of organized crime. Its crimes range from those affecting the banking sector and real estate markets down to the level of the food markets where citizens shop. St. Petersburg, as the second largest city of Russia, shares some of the same criminological attributes. Its crime patterns differ because it has much less capital than Moscow, and, as an important northern port, it shares some crime characteristics with Vladivostok.

Although many contend that the crime situation in Moscow has stabilized as different crime groups have established control over particular regions of the city, the massive violence noted at the end of 2001 suggests that territorial and economic disputes are still a source of significant violence. In the first days of December 2001, Moscow surpassed all previous records for this time period of organized-crime-related killings as nine people were killed in gang shoot-outs, contract killings, and armed attacks.[36] Organized crime and political conflict contribute to the high overall levels of violence in Russia today.[37]

Moscow has very significant levels of money laundering. The concentration of the lightly regulated banking sector and the hundreds of casinos and money exchange offices permit illicit Russian money and illegitimate money from abroad to be laundered with little chance of detection. The lack of regulation of financial markets such as stock market and other investment funds has facilitated the criminalization of these sphere of the economy.

The value of the Moscow and St. Petersburg real estate markets has contributed to the corruption of these markets. High-level officials in these cities demand significant bribes even to learn about the availability of property for rent and purchase. The Mabetex case in Switzerland revealed the very high-level corruption that reached even to Yeltsin's family. In this case, members of Yeltsin's family and entourage had significant amounts paid into their Swiss bank accounts. In return, Mabetex obtained lucrative contracts for the restoration of the Kremlin.[38]

The funds to pay for Moscow's reconstruction came at the cost of the redevelopment of other areas of Russia. According to audits of the Chamber of Accounts (the oversight body of the Russian government), funds intended to reconstruct the gold fields in Siberia and support other infrastructure projects were diverted from the accounts in Moscow-based banks through which these regional transfers were made. The system of Kremlin-favored banks permitted the massive diversion of resources, a practice that Putin has tried to stem by bypassing the Moscow-based banking system and making direct transfers from the treasury to the regions.

Organized crime assumed a significant role in the real estate markets of Moscow and St. Petersburg, particularly in the period of privatization of apartments to their tenants.[39] In the absence of adequate safeguards to protect the elderly, the infirm, and alcoholics, many citizens lost their apartments and simply disappeared with no protection from legal authorities. Organized crime was behind many of the deaths of alcoholics and the elderly who froze, had accidents, or died under mysterious circumstances, freeing their apartments for sale and development. In St. Petersburg, as development has spread to the countryside, the same phenomena occurred with the owners of desirable dachas. Law enforcement has broken almost none of the crime rings operating in this area.

The farmers' markets have been controlled by non-Russian organized-crime groups since the *perestroika* period. The products and sellers traditionally came from outside Russia. Therefore, as the state lost control over the markets and large-scale public markets sprang up on Moscow's outskirts, organized crime groups from the Caucasus assumed control over their activities. These Caucasian-organized crime groups are of mixed nationality, and Armenians and Azerbaijanis who may be at war on their home territory work effectively together in this arena. Conflicts among organized-crime groups have not ended. At the end of 2001, the Baku-born head of the Moscow kolkhoz market was killed. This was the seventeenth murder connected with this market in the past decade.[40]

The Caucasian groups control markets not only in Moscow and St. Petersburg but in Siberia and the Far East as well. This is a historical legacy of their distribution role in the Soviet period when so much agricultural activity became rooted in the second economy.

Ethnic groups are an important part of much organized crime activity in Moscow. They are most present in the illicit sectors of organized crime—auto-theft rings, drug and human trafficking, and arms smuggling. They are also involved in real estate, particularly the hotel sector, some of which have become bases for their operations. Those engaged in arms smuggling and human and drug trafficking form alliances with groups from Eastern and Western Europe to promote their activities. Links are established with Latin American groups and different European groups to promote the drug trade.

Conclusion

The Soviet Union had suppressed crime rates, a consequence of its high levels of social control, high rates of incarceration, and controls over personal residences. With the liberalization of the Gorbachev era, fundamental changes occurred in Soviet crime patterns. Crime rates rose rapidly, and organized crime became a formidable actor in the new economy.

Russia's first decade was a period of major transition, which proved traumatic to many of its citizens. Unemployment rose dramatically, particularly among women, the social safety net collapsed, and many citizens lost their life savings through investments in unregulated banks and markets. Organized crime became a visible force in society, not only through its public displays of violence, but through the key role it played in privatization and politics during the transitional period.

The uneven economic development of Russia resulted in highly differentiated crime problems. Moscow, the capital, also became the financial center of Russia, controlling upwards of 80 percent of the country's financial resources. The crime connected with Moscow involved control over the lucrative financial and real estate markets as well as access to the nation's resources that flowed through its banks. Regions such as the Far East, starved for capital because of their high levels of corruption and remoteness from the center, suffered acutely. The criminalization of the region had dire consequences for the citizens and their social well-being. As a consequence, there has been a mass exodus of Russians from the region. The population of the Far East has declined over ten percent in the past decade. At the same time, the region has become the temporary or long-term residence of citizens of Asian countries, in particular China, who find opportunities in the unregulated economy of the region.

In the past decade, Russian-speaking organized crime has become a powerful force in the international illicit economy. With its valuable natural resources, the myriad ways through which it could launder money through its unregulated financial system, and the high-level corruption of the government, crime groups could operate with impunity at home. Furthermore, they could also protect themselves against foreign prosecutions by pressuring their own law enforcement not to cooperate in international investigations.

Russia has proved accommodating not only for domestic crime groups but for foreign crime groups as well. The pervasiveness of casinos has permitted Latin American and Italian crime groups to launder their money. Crime groups from Asia are visibly present in significant numbers in Siberia and the Far East. Human-trafficking rings and drug operations stretch across Russian territory as the traffickers move their human commodities and narcotics to Western markets.

Even after a decade of transition, the crime rates of Russia remain extraordinarily high, a consequence of ordinary acts of violence combined with the deliberate killings of organized crime. The violent crime rate far exceeds that of western Europe and the United States and more closely resembles that of a country like Colombia that is in the midst of a civil war and where drug groups exercise control in many parts of the country.

The crime problems identified in this paper reveal that crime in Russia is not a peripheral social problem but a central problem of its development. Analyzing Russia's development in terms of its crime statistics and patterns of organized crime reveals that the crime problem is not improving after the years of the initial transition. Although President Putin has pursued some of the oligarchs and emphasized the need for more rule of law, there has been no fundamental change in the crime phenomenon or in the nature of law enforcement. The crime and enforcement problems identified here appear to be structural problems of the Russian state that cannot be addressed easily or effectively in the existing political system. They will influence Russia's political, social, and economic development for many years in the future.

Notes

1. For a discussion of the rise of the Mafia, see Pino Arlacchi, *Mafia Business: The Mafia Ethic and the Spirit of Capitalism* (London: Verso, 1986); Raimondo Catanzaro, *Men of Respect: A Social History of the Sicilian Mafia* (New York: Free Press, 1992); and Henner Hess, *Mafia and Mafiosi: Origin, Power and Myth* (New York: New York University Press, 1998).

2. Louise I. Shelley "Criminal Kaleidoscope: The Diversification and Adaptation of Criminal Activities in the Soviet Successor States," *European Journal of Crime, Criminal Law and Criminal Justice*, vol. 3, no. 3 (1996), pp. 243–56.

3. See N.A. Lopashenko, "Problemy begstva kapitala iz rossii i puti ego vozvrashcheniia," Saratov 2002, research done under grant program of Moscow organized Crime Study Center, available at www.american.edu/traccc; and Alexander Boulatov, "Capital Flight from Russia," paper presented at the Kennan Institute, Washington, D.C., June 4, 2001.

4. Please see the website www.american.edu/traccc for more information.

5. See research of Yuri Andrienko and Louise Shelley, "Crime and Political Violence in Russia," prepared for Workshop on Case Studies of Civil War, Yale University, April 12–15, 2002.

6. V.V. Luneev, *Prestupnost' XX veka.* (Moscow: Norma, 1997), pp. 92–96.

7. N.F. Kuznetsova and G.M. Minkovskii, *Kriminologiia: Uchebnik* (Moscow: Vek, 1998), p. 549.

8. L.K. Saviuk, *Pravovaia Statistika: Uchebnik* (Moscow: Iurist, 1999), p. 382.

9. Louise I. Shelley, "Interpersonal Violence in the Soviet Union," *Violence, Aggression and Terrorism*, vol. 1, no. 2 (1987), pp. 41–67.

10. Kuznetsova and Minkovskii, *Kriminologiia: Uchebnik*, p. 553.

11. Russian homicide figure derived from Andrienko and Shelley and comparative figures from *International Journal of Epidemiology*, vol. 27 (1998), p. 216; and U.S. data are from the *FBI Uniform Crime Report*.

12. V. Luneev, "Geografiia organizovannoi prestupnosti i korruptsii v Rossii (1997–1999)," *Organizovannia prestupnost i korruptsiia: Regional'nyi srez*, vol. 3 (Ekaterinburg: Zertsale-Ural, 2000), pp. 5–15.

13. See Louise Shelley, "Crime and Corruption," in *Developments in Russian Politics*, ed. by Stephen White, Alex Pravda, and Zvi Gitelman (Houndsmills: Palgrave, 2001), pp. 239–53.

14. See the press reports from November and December 2001 produced by the Organized Crime study centers sponsored by TraCCC in Moscow and Vladivostok. Available at www.american.edu.traccc for Moscow and www.crime.vl.ru/docs/obzor/1201.htm for Vladivostok.

15. Svetlana Glinkina, "Privatizatsiya and Kriminalizatsiya—How Organized Crime Is Hijacking Privatization," *Demokratizatsiya*, vol. 2, no. 3 (1994), pp. 385–91.

16. Frederico Varese, *The Russian Mafia—Private Protection in a New Market Economy* (Oxford: Oxford University Press, 2001), pp. 182–83; Louise Shelley, "The Political-Criminal Nexus: Russian-Ukrainian Case Studies," *Trends in Organized Crime*, vol. 4, no. 3 (1999), pp. 81–107.

17. Larisa Kislinskaia, December Press Survey, Moscow Organized Crime Study Center, 2001, available at www.american.edu/traccc.

18. Center for Strategic and International Studies, *Russian Organized Crime and Corruption Putin's Challenge* (Washington, D.C.: CSIS, 2000), pp. 32–39.

19. *Regioni Rossii: Statisticheskikh sbornikh* (v 2 tomakh) (Moscow: Goskomstat Rossii, 2001), p. 827.

20. Louise Shelley, "Internal Migration and Crime in the Soviet Union," *Canadian Slavonic Papers*, vol. 13, no. 1 (1981), pp. 77–87.

21. V. N. Kudriavtsev, "Osnovnye prichiny regional'nykh razlichii v organizovannoi i korruptsionnoi prestupnosti, *Organizovannia prestupnost i korruptsiia: Regional'nyi srez*, vol. 3, pp. 30–36.

22. A. L. Repetskaia, "Kriminologicheskaia kharkteristika organizovannoi prestupnosti v Sibiri (1994–1999)," ibid., pp. 36–38.

23. Ibid., p. 42.

24. Varese, *The Russian Mafia*, pp. 68–72.

25. See press reports of Irkutsk Organized Crime Study Center for 2001, available at www.isea.ru/sait/rezult/inform/obzory/index.htm.

26. O.A. Popova, "Banditry vs. Armed Robbery by an Organized Crime Group," *Organized Crime and Corruption Watch*, vol. 2, no. 1 (2000), pp. 2–15.

27. Anna Repetskaya, "Irkutsk Organized Crime Press Review," *OC Watch*, vol. 5 (October 1, 1999), p. 16; see www.american.edu/traccc.

28. For a discussion of the development of the situation in Ekaterinburg, see Stephen Handelman, *Comrade Criminal: Russia's New Mafiya* (New Haven: Yale University Press, 1995), pp. 73–92.

29. Irina Belousova, "Organized Crime in Yekaterinburg and the Sverdlovsk Oblast," *Organized Crime and Corruption Watch*, vol. 2, no.1 (2000), pp. 17–24; available at www.american.edu/traccc.

30. Varese, *The Russian Mafia*, p. 183; and the Center for Strategic and International Studies, *Russian Organized Crime and Corruption*, pp. 24–42.

31. See Organized Crime Watch on the website of www.american.edu/traccc, which has publications of the Russian Far East center. For in-depth analysis of crime in the Far East see V. A. Nomokonov, ed., *Organizovannaia prestupnost′: tendentsii, perspektivy bor′by* (Vladivostok, Far East University Press, 1999:); V.A. Nomokonov, *Transnatsional′naia Organizovannia Prestupnost′ Definitsii i realnost′* (Vladivostok: Far East, University Press, 2001); and V.A. Nomokonov and V.I. Shulga, "Murder for Hire as a Manifestation of Organized Crime," *Demokratizatsiya*, vol. 6, no. 4 (1998), pp. 676–80.

32. See Moscow and Vladivostok Crime Centers Press Reports for December 2001, available at www.american.edu.traccc for Moscow and www.crime.vl.ru/docs/obzor/1201.htm for Vladivostok.

33. V. A. Nomokonov, ed., *Organizovannaia prestupnost′: tendentsii, perspektivy bor′by* (Vladivostok: Dalnevostochnogo universiteta, 1998).

34. Ibid.; see also Yana Aminyeva and Viktoria Chernyshova, "Organized Crime in the Far East and Maritime Provinces," *Organized Crime and Corruption Watch*, vol. 2, no. 1 (2000), p. 28; on web at www.american.edu/traccc.

35. L.D. Erokhina, "Seksualnaia ekspolitatsiia zhenshchin i detei v rossii," research paper January 2001 for TraCCC grant (Transnational Crime and Corruption Center) on Sexual Exploitation of Women and Children, funded by USIA.

36. Kislinskaia, December Press Survey, available at www.american.edu.traccc for Moscow.

37. Andrienko and Shelley, "Crime and Political Violence in Russia."

38. Paul Quinn-Judge, "A Family Matter," *Time*, September 20, 1999. Available at www.time.com.

39. Konstantin Dobrynyn, "Moshenichestvo v sfere nedvimosti v Sankt-Peterburge: primaia vzaimosviaz′ s organizovannoi prestupnost′iu i korruputsiei," research grant done for St. Petersburg Crime and Corruption Center, 2002; available at www.jurfak.spb.ru/centers/traCCC/article/default.htm.

40. Kislinskaia, December press survey, available at www.american.edu.traccc for Moscow.

For Further Reading

Stephen Handelman, *Comrade Criminal: Russia's New Mafiya* (New Haven: Yale University Press, 1995).

Phil Williams, ed., *Russian Organized Crime: The New Threat* (London: Frank Cass, 1997).

Victor M. Sergeev, *The Wild East: Crime and Lawlessness in Post-Communist Russia* (Armonk, NY: M.E. Sharpe, 1998).

Paul Khlebnikov, *Godfather of the Kremlin: Boris Berezovsky and the Looting of Russia* (New York: Harcourt, 2000).

David Lane, *The Legacy of State Socialism and the Future of Transformation* (Lanham, MD: Rowman and Littlefield, 2002).

7

Putin's Vertical Challenge: Center-Periphery Relations

Darrell Slider

Of all policy areas, Russia's federal system has seen perhaps the most comprehensive set of policies and institutional changes. Vladimir Putin launched a series of initiatives to restructure relations with the regions from the start of his presidency and has devoted sustained attention to the issue.

The Yeltsin legacy in this area was mixed. On the one hand, Yeltsin succeeded in preventing the disintegration of the Russian republic and pushed Russia in the direction of establishing federalism as the basis for relations between the center and the regions. He pushed for the establishment of a Federation Treaty in March 1992, which was signed by all regional leaders except those of Chechnya and Tatarstan. Over the years the treaty was supplemented by bilateral agreements signed by the federal and regional executives that elaborated relations in a number of areas. The 1993 Russian Constitution further entrenched the principle of federalism, outlining in very general terms the division of powers between the center and regions. Regions and their leaders were brought into national politics through the new Federation Council and in efforts at party formation that preceded elections to the State Duma.

There were also many negative aspects of Yeltsin's policies, however. Yeltsin's political weakness meant that he was forced to make compromises and concessions to republic and regional leaders in return for their support. As Russia was establishing itself as a sovereign state, preventing regions (particularly ethnic republics) from withdrawing was accomplished by allowing them, in Yeltsin's famous charge, to "take all the sovereignty you can swallow." Russia's economic crises from 1992 to 1998 meant that the center failed to provide the funding necessary for federal agencies in the regions. The result was that regions began to take control over functions that no federal system entrusts to regions. The center lost effective control over many federal agencies in the regions, as regional leaders had a decisive say over appointments and financed federal operations from regional budgets. The weakness of the center also allowed regional leaders to flout the constitution and federal legislation without fear of punishment. In the exceptional case of Chechnya, the preservation of the federation came at a very high price—armed conflict, which led to a Russian military defeat in the first Chechen war, the killing of many thousands of innocent civilians, a flood of refugees, and the economic devastation of the republic.

Putin came to power with an agenda designed to bring the regions under central control by establishing an unambiguous "vertical authority" with the Russian president at the top and the governors and republic presidents under his control. Part of this effort was designed to correct Yeltsin's "mistakes" in allowing the balance of power to shift toward the regions. Putin had personal experience with regional policy when he served in Yeltsin's administration as head of the office that investigated corruption in the regions. Of course, the war in Chechnya was not viewed as a mistake by Putin; as Yeltsin's last prime minister and then as acting president, Putin pushed for a more aggressive posture toward Chechen rebels. The new "anti-terrorist campaign" became the main source of his political popularity in the run-up to the March 2000 election.

This chapter attempts to assess Putin's record after two years in office. To what extent has Putin succeeded in bringing the regions under control? The overall conclusion reached here is that in almost every case the initially proclaimed set of goals and policies has been subject to compromise, if implemented at all. The central government remains too weak to fully carry out its policies, while at the same time regional leaders have retained significant levers of power that allowed them to resist new pressures from the center. Moreover, the worst violators appear to have been the most successful in fending off these pressures.

Presidential Representatives and Federal Districts

The tone of the drive to increase control over the regions was set by Putin's decision, early in his presidency, to create seven new administrative districts.[1] The new districts were designed to shift the locus of federal policy toward the regions to a level where the center would have a clear advantage. Under Yeltsin, the effort to monitor and direct the operations of federal agencies in the regions was centered on presidential representatives who since 1991 had been appointed in almost all of Russia's eighty-nine regions. The effectiveness of these agents of the center was uneven, despite repeated efforts by Yeltsin's administration to expand their coordinating role. By the late 1990s, most presidential representatives had, in effect, been "captured" by regional leaders who played an important role in their selection and functioning.

Putin shifted the main level of coordination to large districts where governors and republic presidents had almost no instruments of influence, and he appointed seven presidential representatives (*polpredy*) as heads of these administrative entities. In a confirmation of Putin's intent to use the new structure to restore discipline and hierarchy, five of the men named to these posts had backgrounds in the power ministries (e.g., KGB-FSB, MVD, and the military). A separate set of "federal inspectors" was named by the presidential representatives to oversee implementation in most regions.

The creation of the new districts resulted in corresponding administrative changes in most federal agencies. They formed parallel administrative entities at the federal district level, thus making it possible for the presidential representative to meet with overseers of regional agencies such as the police, tax collectors, prosecutors, the anti-monopoly agency, etc. The amount of authority of this new intermediate level of bureaucracy varied from ministry to ministry, however, and it is clear that considerable discretionary power remained at the regional level—and thus was subject to influence by governors.

The new districts and presidential representatives encountered a number of problems in their efforts to establish control over regional affairs. The system of federal districts was created by a Putin decree and had no place in the Russian constitutional or legal framework. Responsibilities of the *polpredy* were never clearly formulated, and each seems to have developed his own style and set of priorities. In many cases, conflicts emerged not only between governors and *polpredy*, but between government ministers and *polpredy*. After seven months' experience, Putin met with his representatives in December 2000 and admonished them: "Federal districts are not new administrative-territorial units (you and I have spoken of this often) and even less are they subjects of the federation. Therefore there is no need to try to create some sort of district governments from federal ministry and departmental subdivisions. . . . The task of Presidential representatives, their function is to coordinate the activities of federal organs in the regions, not direct them, but coordinate."[2]

The federal districts did not have control over financial flows to the regions and were provided with very limited staff. Since early 2001, even their access to the president was not direct. To the chagrin of the *polpredy*, Alexander Voloshin, the head of the presidential administration, oversaw their activities, and much of their activity was reported to Putin through him. Putin meets with his presidential representatives as a group only twice a year, and individual meetings with *polpredy* are infrequent.

It is also the case that part of the work of the *polpred*'s office was devoted to handling petitions by individuals, most of which concerned housing or pension issues. This is in keeping with the traditional role of Communist Party institutions as the place citizens turned to when complaints to lower levels of government went unanswered.

Putin's political popularity did not transfer automatically to his representatives in the new federal districts. A series of polls conducted in 2001 found that Putin's personal approval rating was around 75 percent, while that of governors was also mostly positive—from 57 percent to 50 percent from January to August 2001. Approval ratings of Putin's new presidential representatives, however, were significantly lower at around 30 percent.[3]

Perhaps the most important function of the presidential representatives is to restore federal control over federal agencies. In this, an especially important role is played by personnel policies. This is an area that is particularly hard for an outsider to judge, since information about appointments for most posts are not publicized (much less, information about the biography and loyalties of the appointees). That said, the record appears to be a mixed one.

A special area for attention by Putin has been the main instruments of law and order—prosecutors and the police. Under Yeltsin, governors were given the power to veto appointments to head the provincial police. An amendment to the law on the police that became law in July 2001 placed full power of appointment in the hands of the president upon recommendation of the head of the MVD. A compromise, though, allowed governors to participate in the MVD collegium when the appointment was made.[4] It appears that the strongest governors and republic presidents continue to exert influence over key appointments in their regions. Yuri Luzhkov, for example, refused to accept the appointment in December 1999 of a new MVD chief for Moscow; in July 2001 Putin relented and appointed a candidate who was acceptable to the mayor. In another case, a regional leader with weaker ties to Putin, Tula Governor Vasilii Starodubtsev, was unable to prevent the center from firing the oblast prosecutor. The prosecutor was suspected of having too close a relationship with the governor.[5]

Paradoxically, the effort to increase central control over the regions took place at the same time as an effort to reduce the size of federal bureaucracy in the regions. In 2000 the number of central ministry staff in the regions was cut by 7,500. At the same time, there has been a substantially larger expansion in the regional/local bureaucracy. According to Goskomstat data, the numbers of regional bureaucrats grew by 31,000 in 2000, which was approximately the same increase as in 1999.[6] This illustrates a trend that portends less, rather than more, centralization of administration.

Bringing Regional Laws in Conformity with Federal Law

The Yeltsin period was one of rapid institutional change, and this "churning" effect was reflected in a chaotic legislative environment. Many areas of constitutional law remained unimplemented because the parliament had not passed corresponding legislation. Regions stepped into the breach and passed their own legislation, some of which was not in line with the constitution or impinged on federal responsibilities. In some cases, regional laws were surpassed by federal legislation but were not subsequently revised.

The presidential representatives and other federal agencies were assigned the massive task of examining tens of thousands of laws and directives that had been issued by regional authorities in the post-independence period. The Russian Ministry of Justice created a registry of over 100,000 regional "normative acts"—which includes laws and other government decisions—and in 2001 reviewed over 52,000.[7] Over 5,000 unconstitutional acts of regions were invalidated between mid-2000 and the end of 2001.[8]

The sheer mass of legislation subject to review meant that the entire process was quickly turned into an exercise of the type that veterans of the Soviet system were very familiar with. It was inevitable that the process would become highly formalistic. The hidden assumption behind this effort was that changes in laws and regulations would have a direct impact on the behavior of regional officials, courts, prosecutors, and others responsible for carrying out and enforcing the laws. Yet the effort to rewrite legal documents took place in the absence of serious reforms of the court system, the police, and the prosecutor's office, and before significant changes in the civil service/bureaucracy had been implemented.

Despite this massive effort to revise the existing base of legislation and regulations, many of the most blatant contradictions with federal laws and the constitution were still unchanged as of early 2002. It was reported at the end of 2001 that in Bashkortostan, 72 percent of the existing laws and decrees violated federal law.[9] Similarly, in Tatarstan there was strong resistance from the republic legislature and executive to efforts to force it to conform to federal law. In Yakutia, officials had been able to ignore the 1995 Law on Local Self-Management. As of 2002 the region still did not have elected local government bodies.[10] The preferred solution of many regional leaders was for the federal center to revise its legislation to conform with *their* laws.

In part, the legal anomalies were a consequence of another legacy of the Yeltsin era: the series of bilateral agreements signed between regions and the center. Between 1992 and 1999, 46 of these agreements were signed, accompanied by over 500 protocols that specified arrangements in greater detail. It was Putin's clear intention to abrogate these agreements entirely or at least bring them into conformity with Russian laws. By late March 2002, Deputy Head of the Presidential Administration Dmitry Kozak announced that 28 agreements had already "effectively been annulled."[11] In almost every case, these were agreements that were voluntarily given up by governors, most likely because the cost of defending them was not worth the benefits. A member of Kozak's commission, presidential adviser Sergei Samoilov, argued further that, based on the 1999 law on the division of powers, any agreement that had not been brought into accord with federal laws would be considered nonbinding by July 2003.[12]

Recent experience has shown that agreements with a number of regions, particularly republics, will be extremely difficult to amend or overturn; they are fiercely defended by regional leaders and legislatures. Bashkiria president Murtaz Rakhimov has refused to allow the 1994 bilateral agreement to be terminated. It gives the republic the right to form its own "system of state authority," to control the disposition of land and natural resources, to conclude international agreements, and to create its own national bank.[13] Similarly, leaders of Tatarstan have resisted any diminution of that republic's special status and are demanding that any abrogation of the agreement be accompanied by special dispensations. Only after the Constitutional Court forced it to act did the Tatarstan State Council begin revising the republic constitution. The version that was passed in March 2002 substantially reduced Tatarstan's claims to sovereignty but continued to refer to the bilateral treaty as a basis for relations with the center. It also contained a clause that "the status of the Republic of Tatarstan cannot be changed without the mutual agreement of the Republic of Tatarstan and the Russian Federation."[14]

Eduard Rossel, governor of the Sverdlovsk region, in February 2002 announced that his region had no intention of abandoning its agreement with the center, arguing that "this document had no expiration date."[15] Here and in many other regions, opposition to abrogating the agreements was based on many of the areas covered in the agreements that were not yet covered by federal legislation (such as land use and control over mineral and energy resources). In the Irkutsk oblast, for example, the bilateral agreements had solved a dispute over the division of ownership shares of the region's electrical power system. It was feared that striking the agreements would threaten the ability of the region to influence energy policy.[16] In Kaliningrad, the economic advantages enjoyed by the region are in part due to the bilateral agreement with the center, and there is no interest in abrogating the treaty.[17]

Political Control over Governors

One of Yeltsin's concessions that was to have the most far-reaching consequences in shifting power from the center to the regions was the decision to allow governors to run for popular election. (Most republics had earlier taken advantage of their superior administrative status to elect presidents.) Popular election gave governors a degree of legitimacy that rivaled that of the president and meant that, in practice, they could not be fired.

Putin set out to change this situation by proposing amendments to the law that would allow him to initiate proceedings to dismiss a governor. Amendments to the law on regional government organs permit the president to punish a governor who issues a decree that contradicts the constitution and Russian laws or is charged with a serious crime. The initial version he proposed gave the president great latitude. After a series of compromises were introduced, the amendments allow dismissal only after warnings and after a court rules that a violation has indeed occurred. The compromises provide opportunities for appeals and allow the official time to reverse or annul a previous decision.[18]

The amendments were passed in July 2000, over the veto of the Federation Council, and went into effect in February 2001. Yet Putin has not even come close to putting them to use. When he has apparently decided to remove regional leaders, he has operated largely behind the scenes, using other methods. A number of regional leaders who clashed with Putin, such as Alexander Rutskoi of Kursk, Mikhail Nikolaev of Yakutia, Ruslan Aushev of Ingushetia, and Nikolai Kondratenko of Krasnodar, were either prevented from running or "voluntarily" withdrew their candidacies for reelection—presumably after pressure orchestrated by the Kremlin. One particularly "difficult" governor, Yevgenii Nazdratenko of Primorski krai, was given another job—head of the federal committee that awards lucrative fishing quotas—in return for resigning as governor. The fact remains, however, that a very large number of governors and republic presidents who are out of favor with Putin are still in power. Even in his home of St. Petersburg, Putin has not been able to dislodge the governor. Vladimir Yakovlev, who easily won reelection in May 2000, is viewed by Putin as an enemy who betrayed their common patron, Anatoly Sobchak.[19]

Putin has sought methods other than firing governors as a way to control political developments within the regions. While not publicly seeking an end to elections of governors, Putin has attempted to expand his opportunities to appoint governors. In early 2002 the Putin-oriented parties that dominate the Duma began efforts to revise election laws in order to allow the president to appoint governors under certain circumstances. The first versions of amendments to the law on the rights of voters would have required a 50 percent turnout and the winner to receive over 50 percent in the first and only round of voting. If no winner emerged, the president would be allowed to appoint the governor. This proposal would have given the president enormous powers of appointment, since relatively few regional elections result in an immediate absolute majority in one round. Turnout is also frequently a problem. (The initial law on voters' rights allowed each region to set minimum turnout percentages.) Of gubernatorial elections held between 1998 and 2002, in thirty-three the turnout was less than 50 percent. Later versions of the amendments proposed two rounds of voting, similar to the system used for Russian presidential elections. If the 50 percent turnout level was not reached, however, Putin would still be allowed to appoint a governor for two years. The candidate chosen would have to be approved by the region's legislature.[20] A further proposed change in election procedures would further erode the role of regional leaders: The Central Election commission has proposed that it be allowed to appoint the chairmen of lower-level election commissions, thus undermining the abilities of governors to use their "administrative resources" to control the outcome of elections.

One of the most effective methods for changing the relationships between the center and the regions would be to prevent the reelection of governors who had entrenched themselves during the Yeltsin period. As Union of Right Forces (SPS) leader Boris Nemtsov put it, "if the governors are permitted to work in the regions for 12–16 years, this means economic stagnation, corruption, arbitrary rule by the bureaucrats, and privatization."[21] The October 1999 law on regional organs of power set a two-term limit for governors, matching that of the Russian president. The law did not make clear, however, when the terms would begin to be counted—before October 1999 or only after. A January 2001 amendment supported by governors employed the latter interpretation, thus opening the way for 69 governors to run for a third term and 17 for a fourth. The SPS put forward another amendment on July 2001 that would have reduced those eligible for a third term from 69 to 10, but this was subsequently vetoed by the Federation Council. Then, in November, supporters of the amendment attempted to override the veto, conducting for the first time in Duma history a secret vote for the measure in order to negate the pressure by governors on deputies. Nevertheless, the measure failed to override the veto, in part because it was not supported by Putin's party, Unity.[22] As a result, many prominent governors and republic presidents will be allowed to remain in their posts for many more years. Aman Tuleev, the populist governor of Kemerovo oblast, for example, will be allowed to run for reelection twice more, allowing him to continue to govern the region until 2011.[23]

The dominance of regional leaders, in most cases, extended to regional legislatures that were elected beginning in the mid-1990s. Since national parties were weak in the regions, in most cases regional legislators were either administrators directly dependent on governors or *"praktiki"*—managers of enterprises or farms that needed favors from the regional government. Few deputies were nominated by or supported by national political parties.

Putin proposed a set of provisions similar to those allowing the dismissal of governors that would enable him to disband a regional legislature that refused to submit to federal laws. Again, however, compromises meant that the law would be almost impossible to use. First, the proposal to disband the legislature requires a court ruling that a violation occurred. The legislature is given six months to correct their legislation, after which the president can issue a warning, after which the assembly has another three months to act. Only then can the president submit to the Duma a draft law on disbanding the legislature, a process that can take up to two more months to reach a vote. Again, the complexity and number of steps that would have to be taken virtually rule out its use.

In a long-term strategy to increase potential levers of control from Moscow over regional legislatures, Putin's team began to push for a change in the type of elections in the regions to include a substantial element of proportional representation by party list. The law on political parties adopted in July 2001 prevented the registration of regional parties, requiring that parties register branch organizations in a minimum of forty-five regions. Proportional voting for at least a portion of regional assemblies could potentially increase the role of the center over regional legislatures—"potentially" because regional leaders may be able to dominate the regional branch of at least one of the federal parties.

Not coincidentally, this new strategy coincided with an intensified effort to create a national political party dominated by Putin. Starting in 2002, United Russia *(Yedinaia Rossiia)*, a newly organized "centrist" party created by merging Unity *(Edinstvo)*, and Fatherland/All Russia (*Otechestvo/Vsia Rossiia*) began to operate in the regions. In many regions there was a struggle over who would dominate the new regional party organs. The party's Central Political Council proposed its own list of preferred candidates to head regional branches, with the clear intention of preventing governors from dominating the regional branches.[24] Regional councils and the heads of regional organizations had to be elected at regional founding conferences, but the ultimate choice of a regional party leader had to be approved by the Central Political Council.[25]

In regions where governors or republic presidents were left outside the party, there was a potential for serious new conflicts with the center. This was already apparent in March 2002, when the first elections to regional assemblies after the formation of United Russia began to take place. In Pskov, Governor Yevgenii Mikhailov left the party that brought him to power, the LDPR, and joined Unity in 2000. In 2002, he was expelled from the local branch of Unified Russia on the eve of legislative elections as a result of his failed attempt to control the new party. His response was to attempt to preserve the local branch of the Unity party under his control and to use the local police to disrupt the campaign of Unified Russia.[26] In Sverdlovsk, where elections will be held for the regional legislature in April 2002, the oblast leadership took similar steps. Governor Eduard Rossel had his own political party, For Our Urals (*Za rodnoi Ural*). Oblast authorities did not stand idly by when United Russia launched its campaign. In March the election commission (still dominated by Rossel's people) mobilized prosecutors to investigate illegal funding practices allegedly employed by United Russia.[27]

The long-term electoral potential of United Russia is still in doubt, as is its usefulness as a tool for enhancing central control over regional politics. One of the basic problems facing Putin is the lack of a cadre pool in the regions who could serve as his favored candidates. Trying to set up regional organizations without the backing of the governor or president would most likely result, at least in the short run, in defeat. Given the advantages possessed by an already existing national party, the Communist Party of the Russian Federation (KPRF), efforts to increase control over regional legislatures through proportional representation could backfire.

Rather than weakening governors within their own regions, Putin by his example has tacitly given the green light for governors and republic leaders to move against their opponents in several key areas. In the area of mass media, Putin's policy toward NTV and TV-6 stimulated a new round of attacks on journalists and independent sources of information in the regions. In Lipetsk, in the period preceding elections in the region, authorities took over the only private television station and one of the few media outlets that had been critical of the governor, Oleg Korolev.[28] A wide variety of measures have been used against journalists in other regions: lawsuits, criminal cases, withholding information through special "accreditation" procedures, closing down sales outlets, as well as the usual tax and economic pressures.[29]

Local government—consisting of mayors, district chiefs of administration, and local assemblies—are another source of potential opposition to governors. The Kozak commission in 2002 is supposed to work toward clarifying the powers of this "third level" of government (though only two of the commission's twenty-two members represent the interests of local authorities). If Putin were to ally himself with mayors, he would find support for limiting the powers of governors. To date, however, there is little evidence that the Kremlin is pursuing such a strategy. Governors and republic leaders, while doing what they could to prevent Putin from bringing them under his vertical of authority, often renewed efforts to strengthen their own "vertical" power by undermining the already weak institutions of local government. The current governor of Primorski krai, Sergei Dar'kin, for example, is attempting to cut the number of municipalities in half and then appoint acting chiefs of administration to each. This would allow the governor to solve his ongoing problems with the mayor of Vladivostok, Yuri Kopylov, who otherwise will remain in office until 2004.[30]

Economic Controls over Regions: The New Budgetary Federalism

The Yeltsin period saw a rapid deterioration in federal finances. Partly this was due to the economic collapse that accompanied the breakup of the Soviet Union and, with it, the country's industrial-manufacturing base. The tax collection system itself was badly in need of reform. By 1997, only about 10 percent of GDP was being collected in taxes. An additional factor behind the weakness of the center was the redistribution of revenue sources to the benefit of the regions. For most of the Yeltsin period, approximately 56 percent of government revenues went to regional budgets. Much of this money was transferred as lump-sum payments that were not used for their intended purpose. Some regions, such as Tatarstan, Bashkortostan, and Yakutia, were granted special budgetary advantages that allowed them to keep a larger share (as much as 100 percent) of taxes collected in their regions. Prior to 1998, the use of budget off-sets and in-kind (barter) payments in lieu of taxes further muddied the financial picture and undermined tax flows to the center.

The first two years of Putin's presidency was marked by a very different economic situation. Rising oil prices and the post-1998 economic recovery led to improved government finances. Putin used this and his political strength to increase the role of the center in allocating regional funds. The efficiency of federal tax flows in the 2001 budget was increased by reassigning all of the value-added tax receipts to the center. For the 2001 budget, the share of the budget turned over to regions was reduced from 50/50 to 45/55 in favor of the center. The 2002 budget further tipped the balance away from the regions to 37/63.

Putin's government continued a policy begun in the last part of the Yeltsin period to improve federal monitoring of the expenditure of federal budget money in the regions. The chief mechanism is the increasing use of offices of the federal treasury (*kaznacheistvo*) to dispense budget money in each region. The Ministry of Finance (Minfin) has principal responsibility for seeing that budget funds are spent for the intended purpose through its control-auditing department (KRU). In this role it has been, and continues to be, ineffective. The regional branches of KRU were, until 2002, financed mostly by the regions. They were well-funded and equipped, but unwilling to examine closely the activities of their paymasters. When the agencies became more independent, their funding mostly disappeared.[31] Other agencies have sought to compete with Minfin for a role in monitoring regional expenditures. The Auditing Chamber, revitalized under the leadership of former prime minister and MVD head Sergei Stepashin, made a bid in February 2002 to bring under his supervision the sixty-seven regional control-accounting chambers. Like the Auditing Chamber, the regional chambers are subordinate to the legislative branch. Currently, though, these agencies have a lower status and are even more dependent on regional leaders than on KRU.[32] Putin created a new Committee on Financial Monitoring in January 2002; rather than strengthening the auditing system, the new agency further divided it and took 300 specialists from the KRU.[33] Despite the importance of auditing activities for improving the effectiveness of central allocations, the Putin administration has provided neither clear leadership over the competing bureaucracies nor sufficient funding.

Another new element to budget federalism was a plan announced at the end of 2001 to declare insolvent regions bankrupt and to put them into a type of "receivership" under the direct control of the center.[34] The practical application of this method would appear to be limited, however, since it is beyond the capability of Minfin to take over financial policy in more than a few regions.

In early 2002, Minfin tried a new approach toward the regions. The first change was a presidential decision to almost double, beginning in December 2001, the salaries of budget-funded state employees and to place most of the burden on the regions. This immediately created difficulties even for the what used to be called "donor" regions. Then the Ministry of Finance opened negotiations with regions over the provision of loans to pay for the salary increases.[35] The first two agreements that resulted were negotiated with Krasnoyarsk and Nizhnii Novgorod, both of which are traditionally donors to the federal budget. In each case Minfin demanded strict conditions on the regions to improve their financial systems.[36] Many regions face a similar problem in paying energy debts; Minfin announced that in return for assistance, regions would have to take the politically unpopular step of cutting subsidies and raising customers' utility bills so that 90 percent of the cost of municipal services would be covered.[37] Financial reforms in the regions are badly needed, but again there are doubts about whether the agreements signed will actually be implemented and whether Minfin or any other agency is capable of monitoring the process.

It is also the case that changes in federal budget policy leave untouched many of the instruments available to governors to allocate assets in their regions in ways that are virtually impenetrable to federal agencies. In almost every region a unique system has developed for forming budgets. A special role is played by off-budget funds which are particularly immune to outside scrutiny.[38] Enterprises in a region are often forced to contribute to these funds. In cases where regional legislatures have investigated the expenditure of these funds, corruption and misallocation are the norm. In Tula, for example, a number of special-purpose funds were found to have be used for side payments to officials or poorly monitored operating expenses (including an unspecified amount to restore a Lenin monument).[39]

Federation Council Reform

The Federation Council (FC), or upper house of the Federal Assembly, was created by Yeltsin as part of the 1993 constitution to provide representation from the regions. At first regions elected two members (many were sitting governors), but during 1996–2001 each region was represented by the chief executive and by the chairman of the regional legislature. The constitution gave the FC an important role in the legislative process, with the right to veto legislation passed by the Duma. Since these were not full-time legislators, the members of the Federation Council depended heavily on recommendations by committees and staff. Frequently, the FC vetoed measures adopted by the Duma that would have threatened the prerogatives of the regions.

Putin's "reform" of the Federation Council was designed to reduce its role as an obstacle to recentralization. Its main element was to change once again the method by which members were chosen. Governors and legislative chairmen would be replaced by full-time members who were designated by governors and legislative chairmen. Much of the commentary at the time of this change assumed that the result would be a chamber made up of members who would be much more easily manipulated by Putin, in part because they would be permanently stationed in Moscow. (Indeed, it was assumed that most would be Muscovites.) It is not at all clear that the newly named representatives will be less willing to defend their regions and resist centralizing tendencies. Government officials who reported to the Federation Council in 2001, as the change in membership was gradually being phased in, argued that the new members were even more assertive in defending regional interests than were their predecessors. It is also the case that governors and regional legislatures have the right to recall their representatives without any explanation (though the speaker has suggested that a change in the law would be proposed to require clear cause).[40] It is apparent that Federation Council members will be expected to use their status to lobby government officials for the interests of their regions.

More reason for speculation that the new FC would be more malleable was the formation in February 2001 of the first faction within the Federation Council, the "Federatsiia" group. It was led by pro-Putin members, mostly from the new contingent of deputies, and it grew to include a solid majority of deputies. Members of the faction agreed to vote in unison on important issues. Yet in

January 2002, when the newly constituted Federation Council met and determined its rules, the decision was made to prohibit any factions of members, and the "Federatsiia" group was disbanded. Another piece of evidence of the increasing pro-presidential nature of the FC was the election in December 2001 of Sergei Mironov to the post of speaker of the FC. Mironov, a leading member of the "Federatsiia" group, had served as deputy speaker of the St. Petersburg city assembly before that body chose him their representative to the Federation Council in June 2001. He worked with Putin in St. Petersburg in 1995–96, when Putin was an aide to Mayor Anatoly Sobchak.[41]

Whatever the direction taken by the new members of the Federation Council, reports of its diminished role are premature. Since the body is now functioning permanently, the size and expertise of its staff can be expected to grow substantially. The newly constituted FC will meet more often. In the past, one session per month was typical. Under the new regulations adopted by the body in February 2002, the FC will meet a minimum of twice a month—and usually much more often. One reason for the change was to allow the FC to meet its constitutional obligation to review laws passed by the Duma within a fourteen-day period. Needless to say, the change gives the FC the opportunity to play a more assertive role in the legislative process. Sergei Mironov has also indicated that the upper house plans to get involved at an earlier stage of drafting legislative, including contacts with ministries that provide a large number of draft laws for consideration by the Duma.

Legislation designed to restructure relations between the center, regions, and localities has been designated a priority sphere for 2002, with a key role being played by the presidential commission headed by Dmitri Kozak. In this area, the Federation Council under Mironov has indicated that it plans to take the initiative in this area away from the Duma, which has traditionally favored limitations on the rights of regions at the expense of the center.

Conclusion

When Putin began his assault on the privileges of the regions, there was an immediate and palpable psychological change in the perceptions of where power lay. In the early period of his presidency, Putin appears to have successfully intimidated governors and republic presidents into submission—perhaps with threats of prosecution for corruption. This led many governors to take preemptive steps to undermine their own powers, in the evident hope that this would allow them to preserve the rest of their powers and maintain good relations with Putin. A number of governors, for example, publicly announced that they favored a change from election to presidential appointment of governors. Even leaders of the republics that deviated the furthest from Russian laws found positive elements in Putin's policies. One of the few to question Putin's new direction was Chuvash president Nikolai Fedorov (a former minister of justice under Yeltsin). He even participated in a legal challenge (filed by the republics of Adygeia and Yakutia) in the Constitutional Court to Putin's amendments to permit the dismissal of governors and disbanding of legislatures.[42]

Increasingly over the course of 2001, however, governors and republic leaders appeared to be willing to speak out against central policies. The initial shock of Putin's more aggressive approach toward regional leaders had worn off and was now tempered by the realization that Putin was very cautious in the steps he took against governors. Also, it appeared that the more popular the leader, the less likely Putin would try to challenge his authority. As a result, it is now much more common for regional leaders to stand up publicly for their regions even if this could lead to a conflict with the center.

Alexander Tkachev of Krasnodar krai has become one of the most outspoken governors. Tkachev has repeatedly threatened to disobey any federal law that would allow the sale of agricultural land in the region. In part the controversy represents a populist (and racist) appeal that stokes fears that Armenians and other non-Russians will buy the best land. In November 2001, Tkachev gave a speech in which he said that introducing a market for land in the region would cause an armed revolt, and that he would be in the frontlines, pitchfork in hand.[43] In March 2002, the Krasnodar legislature, following Tkachev's instructions, began preparations to deport refugees such as Meskhetian Turks, arguing that they were in the province illegally. Tkachev promised the Meskhetians "charter flights to Uzbekistan."[44] There is no evidence that these increasingly abrasive, anticonstitutional outbursts raised alarms in Moscow.

The other change that has become apparent after two years in office is that lobbying by regional leaders still works. Under Yeltsin this was standard procedure: he tended to revert to practices with which he was comfortable as a former Communist Party regional first secretary in the Brezhnev era. There was a strong reliance on personal relationships as an important part of the policy process. Regional leaders, most of whom also were a product of Brezhnev's administrative system, understood that lobbying the president and government officials was the most effective way to get favorable treatment from the center on policy matters and budget allocations.[45]

Among the regional leaders who have had the most access to Putin are those who have most consistently been able to avoid complying with federal dictates. Leaders who have had relatively more frequent contact with Putin include Moscow mayor Yuri Luzhkov (eight one-on-one meetings), Tatarstan's Mintimer Shaimiev (four meetings), Bashkortostan's Rakhimov (three meetings), and North Ossetia's Dzasokhov (three meetings). Between January

2000 and March 2002, Putin had a total of sixty-eight individual conversations (those that were officially reported) with thirty-two different regional leaders.[46] This works out, on average, to just over two meetings per month. This appears to reflect a decision to selectively "reward" key governors with a personal meeting, but to avoid overexposure to the competing demands of regional leaders. Putin's staff is reportedly under orders not to allow more than three such face-to-face meetings per month with regional leaders.

Widely described as a "consolation prize" for governors after changes were made to the Federation Council was the formation of a new body, the State Council (Gossoviet). Like many other Putin initiatives, it has no foundation in law or the Russian constitution. The new body offers another channel for governors' personal involvement in national policy debates, though its role is purely consultative. Perhaps most important for the governors is the State Council's function as a direct channel to Putin. A presidium made up of seven members (one from each federal district, the membership of which changes every year) meets the president monthly, while the full body meets once every quarter. Members of the full body were reported to be particularly pleased when Putin proposed adding an informal buffet dinner with the governors after every session of the full council.[47]

Whenever regions face gubernatorial elections, a very high percentage of candidates for governor attempt to tie themselves to Putin, claiming that he supports their candidacies. This is not simply an effort to share in Putin's popularity with most voters, but also a transparent attempt to demonstrate that a candidate will be able to lobby him effectively in the interests of the region.

An optimistic reading of the above would be that the struggle between Putin (seeking central control) and governors (seeking maximal autonomy) will produce a stalemate that would approximate a federalist system. Limitations in the power of each side in this conflict prevent either from achieving victory and will force compromises that will be institutionalized in law. Unfortunately, the means used by both sides in this struggle almost certainly guarantee a nonoptimal, extralegal solution that is likely to hinder the development of a truly federal, democratic political system. While there are efforts—especially that under the direction of Dmitri Kozak to provide a new legal foundation for federal relations—much of Putin's regional policy takes the form of more traditional, personalized relations or else practices inspired by his first career in the KGB.

Notes

1. On this aspect of Putin's policies, see Nikolai Petrov and Darrell Slider, "Putin and the Regions," in Dale Herspring, ed., *Putin's Russia: Past Imperfect, Future Uncertain* (Lanham, MD: Rowland and Littlefield, 2002).

2. "Vystuplenie Prezidenta Rossiiskoi Federatsii V.V. Putina na soveshchanii s polnomochnymi predstaviteliami Prezidenta Rossiiskoi Federatsii v federal'nykh okrugakh," December 25, 2000, from the official presidential website www.president.kremlin.ru

3. Yuri Levada, "Sotsial'no-politicheskaia situatsiia v Rossii v avguste 2001 g. po dannym oprosov obshchestvennogo mneniia," *Obshchestvo na fone* (O.G.I. Press: Moscow, 2001), pp. 101–110.

4. *Izvestia*, July 13, 2001. Available at www.izvestia.ru.

5. *Nezavisimaia gazeta*, November 3, 2001. Available at www.ng.ru.

6. Vitaly Golovachev, "Bureaucrats Continue to Multiply," *Trud*, May 16, 2001. Translation from RIA Novosti in Johnson's Russia's List. Available at www.cdi.org/russia/johnson.

7. "Miniust nadeetsia na priniatie Dumoi zakona o federal'nykh organakh iustitsii v 2002 godu." Available at www.strana.ru, February 20, 2002.

8. Dmitri Kozak in www.strana.ru February 26, 2002.

9. "Bashkortostan's Violations of Federal Laws Tallied," *RFE/RL Newsline*, December 28, 2001.

10. "Yakutskii parlament tormozit protsess vvedeniia v respubliske organov mestnogo samoupravleniia." Available at www.strana.ru, March 21, 2002.

11. *RFE/RL Newsline*, March 27, 2002.

12. Andrei Yegorov, "K 2003 godu iuridicheski znachimiykh dogovorov mezhdu regionami i tsentrom ne budet." Available at www.strana.ru, March 29, 2002.

13. *Vremia-MN*, February 28, 2002.

14. *Vremia-MN*, March 2, 2002; Yurii Alaev, "Tatarstan vse eshe ne chast' Rossii." Available at www.strana.ru, March 29, 2002; *Kommersant-Daily*, March 30, 2002.

15. "Sverdlovskie vlasti ne narmereny otkazyvat'sia ot razdeleniia polnomochii s federal'nym tsentrom." Available at www.strana.ru, February 28, 2002.

16. *Izvestia*, March 19, 2002. Available at www.Izvestia.ru.

17. *Izvestia*, February 18, 2002. Available at www.Izvestia.ru.

18. Valentina Kalikova, "Gubernator zashchishchen zakonom o ego otstranenii." Available at www.strana.ru, January 10, 2001.

19. On Putin's attitude to Yakovlev, see Vladimir Putin, *First Person* (Public Affairs: New York, 2000), p. 113.

20. *Kommersant-Daily*, March 13, 2002. Available at www.kommersant.ru.

21. Interview on ORT, April 24, 2001, quoted in *BBC Monitoring*. Available at www.monitor.bbc.co.uk.

22. *Kommersant-Daily*, November 15, 2001. Available at www.kommersant.ru.

23. "Aman Tuleev budet pravit' eshche 10 let." Available at www.strana.ru, November 9, 2001.

24. *Vremia-MN*, March 6, 2002.

25. The initial candidates to head regional organizations appeared in "Yedinaia Rossiia" opredelilas's andidatami na rukovodiashchie posty v regionakh." Available at www.strana.ru, February 21, 2002.

26. *Kommersant-Daily*, March 25, 2002. Available at www.kommersant.ru.

27. "Sverdlovskii oblisbirkom prenial bespretsedentnoe reshenie." Available at www. strana.ru, March 22, 2002.

28. *Kommersant-Daily*, August 30, 2001. Available at www.kommersant.ru.

29. *Nezavisimaia gazeta*, February 28, 2002; *Nezavisimaia gazeta*, March 28, 2002; *Kommersant-Daily*, December 26, 2001; *Nezavisimaia gazeta*, September 5, 2001.

Available at www.ng.ru and www.kokmmersant ru.

30. *Nezavisimaia gazeta*, March 27, 2002. Available at www.ng.ru.

31. Interview with the head of KRU in Sverdlovsk oblast, Mikhail Shipulin, "Daite nam material'nye vozmozhnosti, i my navedem poriadok." Available at www.strana.ru, March 25, 2002.

32. *Nezavisimaia gazeta*, February 12, 2002. Available at www.ng.ru.

33. "Kontrol'nyi vystrel." Available at www.strana.ru, March 26, 2002.

34. *Kommersant-Daily*, December 5, 2001. Available at www.kommersant.ru.

35. *Nezavisimaia gazeta*, February 11, 2002. Available at www.ng.ru.

36. "Nizhigorodskaia oblast' zakliuchila s Minfinom soglashenie o finansovoi pomoshchi." Available at www.strana.ru, February 28, 2002. "Minfin postavil zhestkie usloviia Krasnoiarskomu kraiu." Available at www.strana.ru, March 4, 2002.

37. "Gotovnost' sub"ektov RF vyiti na 90%-nuiu oplatu uslug ZhKKh–uslovie vydeleniia regionam pomoshchi Minfina." Available at www.strana.ru, March 26, 2002.

38. *Nezavisimaia gazeta*, January 28, 2002. Available at www.ng.ru.

39. *Izvestia*, July 18, 2001. Available at www.izvestia.ru.

40. Sergei Mironov in www.strana.ru, February 13, 2002; and *Nezavisimaia gazeta*, March 13, 2002. Available at www.ng.ru.

41. Interview with Mironov in *Izvestia*, December 5, 2001. Available at www.izvestia.ru.

42. Fedorov later withdrew from the case, arguing the point was moot since Putin has never attempted to apply the law.

43. Andrei Yegorov, "Ichkeriia na Kubani." Available at www.strana.ru, November 9, 2001.

44. *Izvestia*, March 28, 2002. Available at www.izvestia.ru. The new law authorizing deportation, "O prebyvanii i zhitel'stve na territorii Krasnodarskogo kraiia," was passed by the krai assembly on March 27, 2002, and published in www.strana.ru, March 28, 2002.

45. This point is made in Eugene Huskey, *Presidential Power in Russia* (Armonk, NY: M.E. Sharpe, 1999).

46. These figures are from Sergei Sergievskii, *Nezavisimaia gazeta*, September 27, 2001, with updated information from the presidential web site (under *rabochii grafik*) at www.president.kremlin.ru. The numbers include five telephone conversations reported in the *grafik*. Available at www.ng.ru.

47. *Nezavisimaia gazeta*, August 31, 2001. Available at www.ng.ru.

For Further Reading

Darrell Slider, "Russia's Market-Distorting Federalism," *Post-Soviet Geography and Economics*, vol. 38, no. 8 (October 1997), pp. 445–460.

Nicholas J. Lynn and Alexei V. Novikov, "Refederalizing Russia: Debates on the Idea of Federalism in Russia," *Publius: The Journal of Federalism*, vol. 27, no. 2 (1997), pp. 187–203.

Alfred Stepan, "Russian Federalism in Comparative Perspective," *Post-Soviet Affairs*, vol. 16, no. 2 (2000), pp. 133–176.

Matthew Hyde, "Putin's Federal Reforms and their Implications for Presidential Power in Russia," *Europe-Asia Studies*, vol. 53, no. 5 (2001), pp. 719–743.

8

The Challenge of Civil Society

Christopher Marsh

The speed with which the Soviet state came tumbling down a decade ago left most observers and participants alike astonished. After all, the collapse of the Soviet "evil empire" was not precipitated by a military defeat at the hands of its archenemy the United States. Rather, the Soviet system was toppled relatively easily by a lethal combination of pragmatic elites and disgruntled masses. As we examine the policy challenges that the Russian Federation faces today, we should not forget the fact that the Soviet Union was brought down by members of its own society. If the impenetrable and indestructible Soviet monolith could collapse so rapidly and without warning, what are the prospects for regime stability in the new Russia?

Throughout the 1990s, the political landscape of the new Russia began to take shape. A weak and ineffectual regime led by Boris Yeltsin emerged to occupy the position where the strong and overarching Soviet state had stood. Following the October events and the strong presidency introduced by the 1993 constitution, some feared that Yeltsin had the means to build another authoritarian state. Under Yeltsin's leadership, however, Russia remained a weak state, with the country's greatest worries being corruption, economic dislocation, and challenges to its territorial integrity. While Yeltsin may not have developed into a model democrat, neither did he turn out to be a dictator. By the time he resigned the presidency on New Year's Eve 1999, Yeltsin had led Russia through one of the most turbulent periods in the nation's history, and he had significantly contributed to the stability of the new Russian state.

Following his ascension to the presidency, Vladimir Putin has made great strides in strengthening the Russian state, restoring order, clamping down on corruption, and reeling in the power of the regions. While he is to be commended for putting an end to the chaos and disorder that characterized Yeltsin's Russia, there is another side to the orderly and strong state that Putin has been able to put into place in his short time in office. There is a fine balance between a government that is strong enough to provide effective governance and one that remains accountable to its citizens and whose powers are held in check. As James Madison articulated this point in *Federalist Paper No. 51*, "You must enable the government to control the governed; and in the next place, oblige it to control itself."[1] It is in this balancing act that constitutional democracies exhibit perhaps their greatest advantage; they are able to provide responsive government while checking the power of the state.

One of the most significant challenges Russia faces in the twenty-first century is reaching such an optimal balance of power between state and society and consolidating democratic governance. While the Yeltsin period certainly saw progress made in the development of democracy in Russia, the country's first decade of post-Soviet rule was not accompanied by the establishment of effective governance or an efficacious civil society. As the Putin era continues to take shape, the challenge is for the Russian state to become sufficiently powerful to curb crime, corruption, and fragmentation while not becoming so powerful that the state can encroach upon the liberties of its citizens, a particularly problematic situation in a period when civil society is still in a nascent stage of development and faces numerous obstacles to its maturity.

This chapter traces the evolution of state-society relations in Russia and examine the prospects for the attainment of political security—in terms of reaching an effective balance of power between state and society, one that will provide effective governance while not encroaching upon the liberties of individual citizens. I begin with a brief discussion of theoretical issues relating to the concept of civil society. My analysis of state-society relations in modern Russia begins with an examination of the situation on the eve of the Bolshevik Revolution and traces its evolution through the Soviet era to the rise of civil society during perestroika, finally culminating in the collapse of the Soviet Union. Detailed consideration is then given to the situation in the new Russia, focusing first on citizenship and civic participation. Next, I consider the role of political parties and the media, which act as mediating institutions between state and society. I then explore the extent to which Russian society trusts its leaders and political institutions. Finally, I conclude with a discussion of the likelihood that Russia will find an optimal balance of power between state and society, resulting in a government with the power to provide effective governance and a civil society with the efficacy to resist the state's impulse to dominate over the society it is meant to serve.

Civil Society and Russian State-Society Relations

Civil society can be understood as the relatively autonomous realm that rests between the state and the private sphere. At its core, civil society is composed of a "set of diverse non-governmental institutions which is strong enough to counterbalance the state and, while not preventing the state from fulfilling its role of keeper of the peace and arbitrator between major interests, can nevertheless prevent it from dominating and atomizing the rest of society."[2] Of course, there may be a virtually complete absence of a civil society in a particular state at one time or another, with the private realm becoming indistinguishable from the public sphere, as happens under totalitarian rule. Where private individuals come together to cooperate and work collectively to bring about public goods, however, civil society can be said to exist.

Civil society is closely identified with voluntary and associational activities that take place within society, for this is where private interests are joined together and the public interest is articulated. By working to shape public consciousness about the course and direction of society, civil society functions as a mediator between the individual and the state. In so doing, it also acts as a buffer against the state by checking its power and resisting the impulse to dominate the society it is meant to serve. State-society relations can be explored, therefore, in terms of a balance between the strength and efficacy of civil society and the actions and motives of the state.

Civil society is seen as an integral component of a flourishing democratic polity, as civically engaged citizens join together to articulate their wants and place policy demands on governing institutions.[3] While this topic has not proceeded without considerable debate, dissenting voices mostly question the way in which civic engagement is measured and whether its disappearance has actually occurred, not whether or not it is vital to the existence of democracy.[4] If a civically engaged citizenry and a vibrant associational life are necessary attributes of any democratic polity, civil society is also necessary for a new democracy such as Russia to consolidate democratic governance. Indeed, one of the most significant challenges Russia faces in the twenty-first century is democratic consolidation, and this requires a civil society that will promote an optimal balance of power between state and society.

Civil Society and the Rise and Demise of the Soviet Union

Civil society can be conceived of as "institutional and ideological pluralism, which prevents the establishment of a monopoly of power and truth, and counterbalances those central institutions which, though necessary, might otherwise acquire such monopoly."[5] A monopoly of power and truth, of course, is precisely what the Communist Party and its leadership acquired following the Bolshevik Revolution. This was not due to the lack of a civil society in

Russia at the time. Actually, it is quite the contrary, for it was the state, not civil society, that was too weak. The abdication of Tsar Nicholas II in February 1917 left a power vacuum that was only partly filled by the Provisional Government, which remained relatively unorganized and unpopular as it sought to consolidate its power while simultaneously executing a war that was responsible for the devastating conditions in Russian society. It was precisely one of the better-organized groups from within civil society—the Bolsheviks—that effectively took advantage of the state's weakness and seized the reins of control.

Once in power, the Soviet regime was quite successful at narrowing the gap between state and society, with the private realm essentially eliminated as it was fused to the public realm as the state attempted to bring all aspects of social and private life under its control. Under the ideology of Soviet Marxism-Leninism, even the most private aspects of social life were open to the scrutiny of the state, and all citizens were compelled to show explicit support for the Soviet state. Even civic associations and public organizations, such as soccer clubs and youth organizations, did not exist as autonomous social actors, but rather as agents of the state. Organizations such as the Young Pioneers and Komsomol (with rough Western equivalents of the Cub Scouts and Boy Scouts) were not simply patriotic youth organizations, but tools of the state used to inculcate young Soviets with the values and ideals of the official ideology.

A similar situation existed with the Russian Orthodox Church. Despite the fact that Russia has no history of separation of church and state, churches were relatively autonomous throughout much of Russia's history. The situation changed drastically under the Soviets. With atheism instituted as the state religion, the majority of churches were closed down and many believers were forced underground out of fear of reprisals for their deviant behavior, which included not only church attendance but everything from baptism to wedding ceremonies (marriages were to be secular, not religious, occasions and therefore conducted in special government offices). The churches that were not driven out of existence often became tools of the state themselves, with priests regularly informing on congregants for the Soviet secret police.

Such a history of social institutions in Soviet Russia has led to some rather grim assessments of the potential for Russian civil society in the post-Soviet period. While it is true that totalitarian rule abused the already weak civic traditions of many formerly communist societies,[6] some have even argued that the Soviet regime's systematic attack against civil society is the best explanation for the severe problems of post-Soviet democratization.[7] While the Soviet state's intentional attack on unsanctioned forms of civic association and its control over pseudo-civic institutions is beyond question, the implications of this are not entirely clear and the evidence mixed.

For one, the themes of the state-sponsored, regimented participation of the Soviet era espoused ideals such as civic duty and patriotism,[8] values that are perhaps still useful in post-Soviet Russia. The fact that civic organization was directed by the state is perhaps less significant than the values that were actually promoted. The Soviet-era ideals of modesty, love of country, and working for the common good, however, are in fact being replaced by goals such as career advancement, success, and "acquiring savvy and learning how to play the game."[9] The fact that people are becoming more achievement-oriented and developing values more compatible with a free-market economy has positive implications for civil society as well, for independent economic activity is one of its strongest pillars.

It must also be borne in mind that the Soviet regime was not entirely successful in its attempt to suppress unsanctioned civic activities. A civil society most certainly continued to exist in Russia under Soviet rule. While pseudo-civic activities were dominated by the state, thus denying them their role of serving as truly autonomous societal actors, other unsanctioned activities took place, some with the government's tacit approval, others despite the state's best attempt to prevent them. This included dissident writers who told the true history of the Soviet state, and even those who read and copied their works through the *samizdat* process so that others could read them as well. Strikes by coal miners, participation in underground religious services, and underground dissident organizations espousing causes such as international human rights all took place under the nose of the state.[10] Despite its best efforts to suppress civil society, the state's control over the hearts and minds of the people was never total. The argument can even be made that in this way Russia was able to maintain an alternative democratic political culture throughout the Soviet period.[11] Underground civic activity would also facilitate the reemergence of civil society during the *perestroika* period, when the state began to allow freer expression as part of *glasnost*.[12]

The civil society that emerged following Gorbachev's introduction of the policy of glasnost drew upon the suppressed civic activity that existed during the Soviet Union and combined with the pseudo-civic organizations, such as scientific associations and even stamp-collecting clubs, that were flowering into genuine civic actors and entering the public arena to promote various concerns and place demands on the state for further reforms.[13] Russia's nascent civil society had great trouble defining itself, however, and many of the "civic" organizations that emerged at this time hardly deserved the title. The delay had tremendous costs, as Russia's civil society was not able to develop to its potential.

Although unable to take the lead in democratization efforts due to its relative weakness, Russia's civil society did side with liberalizing forces in the battle against antireformists and hardliners, culminating in the collapse of the Soviet Union. Although civil society most certainly played an important role in the process of Soviet democratization, it would be an exaggeration to conclude that it wrested the reins of power from the Soviet regime. The fact is that the Soviet state relaxed its grip on society as time went on, finally liberalizing under Gorbachev and his policies of glasnost and perestroika. As with the power vacuum created by the weakening of the Provisional Government in 1917, it was the relaxation of state control in the late 1980s that allowed civil society to reemerge and perform its rightful function as a buffer between the state and private spheres. While the democratization of the Soviet Union was certainly an example of "democratization from above," civil society played a critical role.

Civil society's most difficult tasks lie ahead, however, including the formation of a liberal national consciousness and the exercising of the liberties and freedoms it fought so hard to attain. Democracy can only be built upon a firm foundation of civil society. Absent that, the state may give in to the eternal impulse to silence opposition and to become the sole voice of society, thus reverting to Russia's authoritarian past in the process. But is Russian civil society up to the tasks that lie before it?

Civil Society in the New Russia

The civic enthusiasm that burst forth and spilled onto the streets of major Russian cities during perestroika turned out to be mostly fleeting. During the 1990s, Russians seemed to withdraw from politics to concentrate on economic survival. Cooperative efforts at managing social problems continued to exist, however, including activities such as the Spring Beautification Day and citizen street patrols (*druzhiny*) for high-crime areas in various cities.

It has been said that history predisposes Russians toward mutual mistrust and toward stoic acceptance of what the government does rather than self-confident influence over it.[14] Russians' attitudes toward civic life may themselves be an obstacle that must be overcome for democracy to take root. Political interests still exist, however, perhaps now more so than ever before, and citizens have a stake in what the government does. But has this translated into a vibrant civil society? While the streets are no longer filled with enthusiastic and outspoken supporters of change, such people are a rare sight even in the most established democracies. What is needed are citizens, not necessarily protesters or revolutionaries.

A great deal of the responsibility for the success of democracy rests with the people themselves. To serve as an effective check on the power of the state, Russia's citizens need to be interested, active, and politically efficacious. Engaged citizens must not only follow events and participate in formal political and informal civic activities, they must also make informed choices about the country's future. In developing such a citizenry, Russia today faces formidable obstacles, including the legacy of Communist rule, economic turmoil, and voter apathy.

It appears that Russians have not lost their interest in politics. While politics may rank last in their hierarchy of values, behind family and economic well-being, fully 38 percent of Russians say that politics is important to them.[15] The fact that only a little more than one-third of the population exhibits strong feelings of political efficacy should not be cause for alarm. After all, even the civic political cultures of the most democratic countries are based on a combination of civic and uncivic elements.[16] This segment provides a substantial basis upon which democracy can be built and should promote the further growth of Russian civil society.

Based upon the percentage of Russia's eligible citizenry that participates in elections, the degree of civic engagement is rather strong. Electoral turnout in Russia has been consistently high since the introduction of competitive elections in 1989, when it reached almost 90 percent for the elections to the Soviet Union's new legislature, the Congress of People's Deputies. Although turnout has declined since then, it has remained near 60 percent in all subsequent national elections, except for a dip to slightly more than 50 percent in the 1993 Duma elections, which were conducted hastily in the wake of the forced dissolution of the Soviet parliament. For presidential elections, which are arguably the most significant, turnout has consistently been near 70 percent. This stands in sharp contrast to presidential elections in countries such as the United States, where turnout hovers only around 50 percent. The high level of political participation is not limited to national elections, moreover, as turnout for local and regional elections is quite high in many corners of the country as well.

In an attempt to increase citizen participation in the political process, government agencies and domestic and foreign civic organizations have sponsored numerous programs, such as civic awareness initiatives and "get out the vote" promotions. Perhaps the most innovative approach has been the use of voting "lotteries" in several districts, where voters are entered into drawings for various prizes. In the fall of 2001, the government even sponsored a civil society conference, inviting organizations from across the country to Moscow to take part. Although voting lotteries may effectively increase voter turnout, it does not increase the efficacy of Russian citizens, since they are not participating in the political process for the good of the country but for a chance at their own reward. Likewise, civil society conferences may promote the development of associations in Russia, but the state *sponsoring* civil society is a contradiction in terms and probably does more harm than good in developing a genuine civil society.

In less direct ways, citizens are informed about political issues and mobilized for political action through civic participation. In this sense, participation in voluntary associations, such as fraternal organizations, sporting clubs, and cultural associations, is an important function of civil society. Although the absolute number of clubs and cultural associations in Russia declined by slightly more than 10 percent during the 1990s,[17] this decline is attributable to the transformation of such organizations from regimented participation under the Soviet regime to genuine voluntary participation in the new Russia. The loss of state sponsorship was a death blow to many such institutions, while the ones that survived were based not only on genuine support for their civic activities, but on the financial assistance necessary to keep them alive as well. In addition, many new voluntary associations sprang forth to fill the void left by the numerous state-sponsored organizations that died off.

At the close of the millennium, nearly one-third of all Russians belonged to voluntary associations.[18] Moreover, 33 percent had performed voluntary work "for the good of the homeland," while more than 40 percent are selflessly willing to do what they can to help build Russia's civil society.[19] This represents a huge amount of untapped civic enthusiasm. These same people, however, are reluctant to participate in voluntary activities organized by organizations such as political parties, public associations, and even the church, due at least in part to the negative connotations associated with these organizations as part of the Soviet legacy.

This resource may not lay idle for long, however, as trust in these institutions has risen substantially in the past few years. Trust in the Russian Orthodox Church was at an all-time low in the early 1990s due to the church's complicity in charging money for food and clothing that had been provided to the church free of charge to be distributed as aid. Due to this fact, along with the church's problematic relationship with the Soviet regime, it took quite some time for Russians to regain trust in this social institution. Between 1999 and 2001 alone, however, trust in the Orthodox Church increased from just over 56 percent to 66 percent, with the increase almost entirely attributable to a rise in those who completely trust the church.[20] During the same period the percentage of Russians who did not trust the church at all declined from 15 percent in 1999 to about 10 percent in 2001.[21] While it seems that social institutions such as the Orthodox Church certainly have a legacy to overcome, it also seems that with the passing of time and successful reform they can regain their role as integral components of civil society.

The jury is still out, however, on whether the church is capable of renewal and innovation, and whether it would seek to come to terms with authoritarian forces were they to gain strength in Russia.[22] Although we cannot be certain of the answers to such questions, the church certainly plays a critical role in state-society relations in contemporary Russia.

There is another side to civil society besides that of political participation and social organization. After all, one of the most crucial ways civil society can check the power of the state is by organizing rapidly in defense of a cause, whether it be in the form of public outcry, strikes, or even spontaneous demonstrations. That civil society may lie dormant until collective action is necessary makes any determination of its strength highly problematical.

In this regard, Russians have also shown the strength of civil society by taking to the streets during the August coup in 1991 and participating in the October events of 1993. While there was perhaps as much apathy to these events as there was active resistance and participation, Russians did not sit idly by as their fates were determined by others. In the absence of such tumultuous events, Russians have also continued to organize to achieve collective action. Following the murder of popular democrat Galina Starovoitova in November 1998, crowds gathered for demonstrations and memorial services, with much of the country even observing a "lights out" in her memory. More recently, in April 2001 between 8,000 and 15,000 Russians assembled on Moscow's Pushkin Square to protest government actions aimed at limiting freedom of speech.[23] These and numerous other activities that frequently occur throughout Russia attest to the ability of Russian citizens to place demands on the government instead of apathetically sitting by as deferential subjects. Although on a daily basis it may seem as though Russia lacks an efficacious civil society, one seems to emerge in times of need.

While it may appear that civil society in post-Soviet Russia suffered a drastic decline, evidence also exists to attest to the vibrancy of associational life in Russia and the relatively high level of political participation. At the least, we can conclude that it is not entirely absent and that Russians are not as apathetic and apolitical as many assume.[24] This picture of voluntary groups of private citizens engaging in civil society is in stark contrast to that painted by those who feared that the Soviet legacy of state-sponsored, pseudo-civic activities would inhibit the formation of genuine voluntary associations in the post-Soviet period.

Mediating Institutions: Political Parties and the Media

While civic participation keeps citizens informed and politically active, the strength of civil society must be understood as a resource whose potential can lie dormant during periods of relative stability. Even though it is a very powerful resource, it takes a substantial impetus to bring individuals together for collective action. On a daily basis, therefore, individual citizens play less of a direct role in political affairs. They are still vital components of the political process, however, as individual citizens join together to articulate their interests and compel the government to address them. This process is conducted primarily by political parties, which aggregate the interests of the people and seek to direct the state. The media also play important roles, not only keeping the public informed but also keeping a watchful eye on the government and playing a critical role in mobilizing the masses during political crises. Together, political parties and the media act as mediating institutions between state and society, keeping government responsive and accountable to the people.

Political parties lie at the nexus between the state and civil society. Parties aggregate mass interests into political platforms and mobilize the people as they seek to win political power, with independent citizens becoming elected representatives in the process. Once they win representation, parties seek to direct policy and to represent their constituents' interests. In this sense, they are the channel through which government is held accountable and responsive to the people.

During the Soviet period, the Communist Party had a purpose other than the representation of constituent interests. Despite the fact that the Soviet Constitution declared the party the "guiding force of society," party posts were filled from the top down, and therefore the party may have "guided" society but it was not accountable to it. This Constitution granted another benefit to the party—an exclusive monopoly, with no other parties allowed to exist. Since the Soviet Union was to be a "dictatorship of the proletariat," only the party that represented the interests of the proletariat could exist— i.e., the Communist Party. This all changed in March 1990, when the Soviet legislature brought the Communist Party's monopoly on power to an end by amending Article 6 of the Soviet Constitution.

Within a short time, Russia had many parties to choose from. By the time the December 1993 Duma elections were held, more than 100 parties had formed, although only 13 managed to get on the ballot. Within two years, 43 parties registered for the December 1995 Duma elections, including such strange creations as the Beer Lovers' Party. By December 1999, the number of parties had increased to almost 200, although, due to stricter registration procedures, only 26 parties were placed on the ballot.[25]

Such a large number of parties competing to be the voice of the people is problematic in several regards. First, this number of parties is not necessary to represent effectively the various segments of the electorate. Many parties, therefore, end up with an ambiguous platform or emphasize regional or fringe issues, neither of which helps citizens decide which party will best represent their interests. Second, numerous parties vying to represent the same segment of the electorate results in vote splitting and diminishes the strength of their political voice. This phenomenon has been observed across the globe. Wherever multiple parties or candidates compete for the votes of the same segment of the electorate, a smaller but unified segment of the electorate may prevail at the polls despite their lower support numerically.[26] In Russia, we have seen that giving the electorate such a large choice results in parties splitting the votes of those who wish to support a particular political agenda.[27]

To promote party development and consolidation, in 2001 the government passed a new law on political parties that provides greater incentives for large parties to form and makes things difficult for smaller parties to survive. The law limits participation in elections to duly registered parties and institutes strict rules and procedures for organizations seeking party status.[28] Although these changes are in many ways positive, they also have the potential to diminish the autonomy of political parties and therefore could derail Russia's path to democracy. After all, to remain responsive to society, parties must be free to form and compete in elections.

The new party law may already be having its desired effect. In November 2001, the two major centrist parties—Unity and Fatherland-All Russia (FAR)—merged to form United Russia. This merger of the two largest centrist parties remedies the situation that resulted when the two competed against each other in the 1999 Duma elections, allowing the Communist Party to receive more votes than any other single party (24.29 percent). This occurred despite the fact that together Unity and FAR, which took second and third place, respectively, received a combined total of 36.64 percent of the vote (Unity 23.32 percent; FAR 13.32 percent). If elections were held again, however, there is no guarantee that things would work out to the new party's advantage. A recent survey found that only 15 percent of those polled would vote for United Russia, while 21 percent would vote for the Communist Party.[29]

The media is also a critical component of civil society. It is the public's number one source of information, keeping citizens informed and educating them on political issues. To fulfill that role, it monitors government activity and provides society with information on what the government is doing. The press thus functions as the eyes and ears of the public, promoting openness in both political and economic affairs in order to prevent any dangerous concentration of power that might threaten civil liberties.[30] When crises arise, the public usually learns about them through the media, and thus it plays a key role in mobilizing the masses in periods of crisis. To perform its role as a mediating institution, it must remain free from state domination and serve the best interests of society. On both accounts, the media in Russia today is in an uncertain position.

During Soviet times, the media were directly controlled by the state, and while they kept the public informed, it was simply a matter of the state disseminating information and political propaganda to the people. The media could not perform its function of governmental watchdog. This began to change under Gorbachev and his policy of glasnost, as the media were allowed to publicize events that previously had been kept secret. Eventually, the floodgates opened and independent newspapers and television and radio stations emerged.

Under Yeltsin, the media flourished in terms of their ability to publish and criticize, although many media groups ran into financial difficulties and their criticism bordered on yellow journalism. It was also not as free as it could have been, since they were controlled by a small number of business tycoons, known as oligarchs, who were able to amass tremendous financial resources during the transition from Communism. Despite their negative points, Yeltsin found common ground with the oligarchs since they both shared the same greatest fear—a return of the Communists to power. The oligarchs and their media empires thus supported Yeltsin's electoral victories.

The media also played an important role in Putin's election campaign. As soon as he got into office, however, Putin launched an attack on the media and the oligarchs who control them. Vladimir Gusinsky and his Media-Most empire wound up in open opposition to Putin's regime and was almost demolished in April 2001. The takeover of NTV and the assaults on *Segodnia* and *Itogi* were only the beginning (the pretext to the freedom of speech rally mentioned above).[31] In January 2002, the last independent national television channel in Russia, TV-6, was closed down, greatly diminishing the voice of an independent media.

It is difficult to discern which is Putin's target: the independent media or simply the oligarchs who run them. Despite outspoken opposition in the West to the attack on the free press in Russia, which may sour U.S.-Russian relations, Russians do not seem as worried. In March 2000, 55 percent of Russians polled said that they expected and hoped that Putin would strengthen state control over the media.[32] Regarding the closing of TV-6, only 7 percent of Russians think Putin played a major role in this affair, considering it instead a conflict among "economic actors."[33]

In one regard, the loss of the independent media in Russia might not be such a loss after all, as Russian citizens do not seem to rely upon it to a great extent. Only 13 percent said they pay attention to it during electoral campaigns, while only about 10 percent rely upon newspapers or privately run

TV stations. The largest response among those polled—43 percent—was that they rely primarily on their own experience.[34] This could be another legacy of the Soviet system, when all media were state-run and merely propaganda machines.

While it is probably not beneficial for Russian society that the state has so much control over the media, its control by the oligarchs probably was not the answer either. The decline of media freedom, however, will impede Russia's ability to govern itself and to oversee its leaders and restrain them from imposing themselves upon society.[35]

Does Society Trust the State?

Perhaps no other aspect of state-society relations cuts to the center of the issue more deeply than the extent to which society trusts and supports the state that governs it. This is an important issue in Russia, with the relations between state and society meant to interact according to the cultural ideal of *sobornost*, which can be loosely translated as "conciliarism" or "togetherness." At the center of this concept is the desire that the various segments of society function and interact with the state in harmony, and the highest expression of this ideal would be to have a leader that everyone supports. While the development of a harmonious relationship between state and society in Russia is in many ways positive, the emphasis on harmony between ruler and ruled may not be conducive to creating citizens out of subjects.

This ideal was not met throughout the 1990s, however, since Boris Yeltsin's support was consistently very low, with his approval rating hovering in the single digits for much of the decade. In fact, he was able to win reelection in 1996 only with a concerted effort by the media—which smeared the other candidates as much as it supported Yeltsin—and a multimillion-dollar reelection campaign.

President Vladimir Putin has come closer by far than any other politician in post-Soviet Russia to fulfilling this ideal. From his comments during his inaugural address, it even appears that this is his aspiration, as he remarked, "I appeal also to those who voted for other candidates. I am convinced that you voted for our common future, for a better life, for a flourishing and strong Russia. Each one of us has his own experience, his own views, but we must be together. We have to do a great many things together."[36] More recently, Putin has said that "the greatest achievement of the recent past is the stabilization and the reaching of a certain consensus in society," a consensus that "is helping to implement decisions that are vital from the standpoint of modernizing our country's political system and economy."[37] Not only does Putin appeal to the people, Russian citizens have great faith in their president. Two full years after his election, over 70 percent of Russians continue to express faith in Vladimir Putin.[38] This stands in sharp contrast to the Yeltsin years, when the president's approval rating hovered in the single digits.

Russians not only trust their chosen leader, they also trust the country's political system. In October 2001, 60 percent of Russians polled expressed faith in the State Duma, while 72 percent trusted the armed forces. And despite its checkered past, 51 percent of Russians trusted the Federal Security Service (the successor to the KGB). Finally, with almost 85 percent of Russians expressing faith in the Ministry of Emergency Situations, Russians have more faith in this body than in any other political institution—even more than in Putin himself.[39]

These facts are relevant to our discussion of state-society relations in Russia for several reasons. First, for Russia to remain stable and to continue along the path to democracy, society must express its trust in their leaders and state institutions. Second, underlying the issue of trust is the legitimacy of the system itself. Any country's legitimacy must derive from the citizens' willing cession of political authority to the state, which makes authoritative decisions for society. The high levels of trust expressed in the legislature, military, and Putin himself all suggest that the vast majority of Russians consider the state and its institutions legitimate. Unlike the Soviet regime, which took several years to consolidate its power and whose legitimacy was perhaps always in question, the government of the Russian Federation today is overwhelmingly accepted as the legitimate political authority and its stability seems secure.

The Challenge of Civil Society

Russia has come a long way over the past few decades, but in particular over the past few years. Today, a civil society exists in Russia and it seeks to challenge the overarching tendencies of the state, if only on occasion. Whether or not civil society challenges the state to a sufficient degree, however, is still open to debate. The oligarchs remain in opposition, but they have been forced either to flee the country or to reach an accommodation with Putin. Other potential actors that could restrain Putin's actions have also fallen in line. Regional leaders who battled it out against Yeltsin in the 1990s are even lining up to align with Putin. A troubling example of this was seen during Putin's electoral campaign in winter 2000, when he received the support of many governors and republic presidents despite the fact that he was simultaneously launching an attack against their privileges as part of his policy of strengthening vertical power in the country. Not only did these regional leaders rally their populations behind Putin, they may even have stuffed a few ballot boxes along the way as a sign of support. As strange as it may sound, actions such as these are perhaps additional manifestations of the Russian cultural ideal of *sobornost*, albeit ones that may actually stand in the way of democracy.

The tremendous support for President Putin is certainly a sign of unity, but is it contingent upon his benevolent leadership? Were he to backtrack on the road to democracy, would society follow him? When asked if they would allow the state to temporarily limit freedom of speech and democratic elections in order to restore Russia's stature as a great power, 52 percent said they would not while 36 percent responded affirmatively.[40] These numbers may not be as promising as they might first appear. While a bare majority is opposed to such measures, the fact that more than one-third of the population would potentially support the suspension of these critical democratic rights is very troubling. Although some might try to take comfort in the fact that a majority would oppose such moves, we do not know what percentage of the population would be willing to fight to preserve these rights. This issue is of critical importance because if these rights were limited even on a temporary basis, once instituted, civil society would be greatly hindered and there would be virtually no incentive for the state to reestablish them again.

One positive sign is that Putin seems to be very responsive to public opinion, even to the point of criticism.[41] Perhaps he also strives for the harmony of *sobornost*, and wants to be responsive to the people. This would bode well for state-society relations, because Russians do not want to give up their freedoms and democracy. But is public opinion enough to hold Putin's ambitions in check? While Russia under Putin certainly seems to be making progress in developing a state with the power to provide effective governance, the challenge remains for the country to develop a civil society with the efficacy to resist the state's impulse to dominate the society it is meant to serve.

There are two distinct challenges of civil society referred to in the title of this chapter. First and foremost, the challenge of civil society refers to the complex issues that confound the development of an efficacious citizenry and vibrant civil society in Russia. This includes the development of trust in one's fellow citizens and the willingness to give of one's time and resources for public goods at a time when most people are struggling just to make ends meet. Secondly, civil society must not only compel the state to provide effective governance, it must also pose a challenge to the state's impulse to dominate the society it is meant to serve. This means that if Putin's effort to restore vertical power in Russia does not end with curbing corruption of regional leaders, but extends to the suppression of legitimate dissent and the open discussion of ideas, Russian civil society cannot retain its trust in its beloved leader but must organize to resist the imposition of another authoritarian regime. The development of civil society, the attainment of an effective balance of power between state and society, and ultimately the consolidation of democracy are thus critical challenges for Russia in the twenty-first century.

Notes

1. Alexander Hamilton, John Jay, and James Madison, *The Federalist Papers* (New York: Modern Library, 1937), p. 337.

2. Ernest Gellner, *Conditions of Liberty: Civil Society and Its Rivals* (New York: Penguin, 1994), p. 5.

3. For strong arguments for the necessity of civil society for democracy, see Robert Putnam, *Making Democracy Work: Civic Traditions in Modern Italy* (Princeton: Princeton University Press, 1993); and Robert Putnam, *Bowling Alone: The Collapse and Revival of American Community* (New York: Touchstone, 2000). For a balanced account of the role of civic engagement in American democracy, see Theda Skocpol and Morris Fiorina, *Civic Engagement in American Democracy* (Washington, DC: Brookings Institution Press and the Russell Sage Foundation, 1999).

4. For a strong counterargument to Putnam's "Bowling Alone" thesis, see Everett C. Ladd, "The Data Just Don't Show Erosion of America's 'Social Capital,'" *Public Perspective*, vol. 7, no. 4 (1996), pp. 1, 5–22.

5. Gellner, *Conditions of Liberty*, pp. 3–4.

6. Putnam, *Making Democracy Work*, pp. 183–184.

7. Thomas Nichols, "Russian Democracy and Social Capital," *Social Science Information*, vol. 35, no. 4 (1996), p. 631.

8. Samuel Harper, *Civic Training in Soviet Russia* (Chicago: University of Chicago Press, 1929).

9. Yuri Levada, "People and Their Values: Hopes Associated with Perestroika Have Given Way to Nostalgia for the Past and Efforts to Adapt to New Realities," *Obshchaia Gazeta*, No. 20, March 20–26, 1999, p. 5. From the *Current Digest of the Post-Soviet Press*, vol. 51 (June 23, 1999), p. 12 [henceforth, *CDPSP*].

10. Ludmilla Alexeeva, *Soviet Dissent: Contemporary Movements for National, Religious, and Human Rights* (Middletown, CT: Wesleyan University Press, 1987).

11. Nicolai Petro, *The Rebirth of Russian Democracy: An Interpretation of Political Culture* (Cambridge: Harvard University Press, 1995).

12. Mary Buckley, *Redefining Russian Society and Polity* (Boulder: Westview Press, 1993).

13. Nicolai Petro, "Perestroika from Below: Voluntary Socio-Political Associations in the RSFSR," in *Perestroika at the Crossroads*, eds. Alfred Rieber and Alvin Rubinstein (Armonk, NY: M.E. Sharpe, 1991), pp. 102–135.

14. Timothy Colton, *Moscow: Governing the Socialist Metropolis* (Cambridge: Belknap Press 1995), pp. 748–749.

15. Elena Bashkirova, *Value Change and Survival of Democracy in Russia, 1995–2000* (Moscow: ROMIR, 2001). Available at: www.romir.ru.

16. Gabriel Almond and Sydney Verba, *The Civic Culture: Political Attitudes and Democracy in Five Nations* (Princeton: Princeton University Press, 1963).

17. From 66,000 in 1992 to 58,600 in 1996. *Rossiia v tsifrakh, 1998* (Moscow: Goskomstat Rossii, 1998), p. 170.

18. Bashkirova, *Value Change and Survival of Democracy in Russia.*

19. Grigory Kakovkin, "Bricks and Mortar of a Civil Society," *Izvestia*, May 22, 1999, p. 1. From *CDPSP*, vol. 51 (June 23, 1999), p. 12.

20. The figure for total trust in the church is determined by combining those that completely agree and those that somewhat agree that the Russian Orthodox Church is one of the most important social institutions in Russia. ROMIR, "Sotsial'no-Politicheskaia Zhizn' Rossii v Zerkale Obschestvennogo Mneniia," *Informatsionno-analiticheskii Ezhemesiachnii Biulleten,' 2001 god.* Available at: www.romir.ru.

21. Ibid.

22. Wallace Daniel, *Civil Society and the Russian Orthodox Church.* Unpublished manuscript, Baylor University, January 2002, p. 5.

23. Mikhail Berger, "Moscow Has a Lot of Free People," *Segodnia*, April 2, 2001, pp. 1–2. From *CDPSP*, vol. 53 (May 2, 2001), p. 3.

24. For more on the strength of civil society in Russia, see Christopher Marsh and Nikolas Gvosdev, eds., *Civil Society and the Search for Justice in Russia* (Lanham, MD: Lexington Books, 2002). For more on how civil society may promote democratization in Russia, see Christopher Marsh, *Making Russian Democracy Work: Social Capital, Economic Development, and Democratization* (Lewiston, NY: Mellen Press, 2000); and Christopher Marsh, "Social Capital and Democracy in Russia," *Communist and Post-Communist Studies*, vol. 33, no. 2 (June 2000), pp. 183–199.

25. Christopher Marsh, *Russia at the Polls: Voters, Elections, and Democratization* (Washington, DC: CQ Press, 2002).

26. Gary Cox, *Making Votes Count: Strategic Coordination in the World's Electoral Systems* (Cambridge: Cambridge University Press, 1997), p. 4.

27. Christopher Marsh, "Civic Community, Communist Support, and Democratization in Russia: The View from Smolensk," *Demokratizatsiia*, vol. 8, no. 4 (Fall 2000), pp. 447–460.

28. Svetlana Lolayeva, "Count Off by Twos!" *Vremia Novostei*, February 8, 2001, p. 2. From *CDPSP*, vol. 53 (March 7, 2001), p. 3.

29. "For Which of the Following Parties or Movements Would You Be Most Likely to Vote if There Were Elections to the State Duma Next Sunday?" VCIOM Survey, February 22–26, 2002. Available at: www.russiavotes.org.

30. Georgy Bovt, "The Russian Press and Civil Society: Freedom of Speech vs. Freedom of Market," in Marsh and Gvosdev, *Civil Society and the Search for Justice in Russia*, pp. 91–104.

31. *Segodnia* is a daily newspaper that was part of the "Media-Most" group that was eventually shut down by the new owner; *Itogi*, a political magazine, remains in operation, but the former staff was purged in a reshuffling of personnel.

32. "Russians Favour Putin Strengthening Control Over Media," VCIOM Survey, March 3–19, 2000. Available at: www.russiavotes.org.

33. "What Do You Think Was the Main Reason for the Closure of TV-6," and "Do You Think Putin Played a Part in Deciding the Fate of TV-6?" VCIOM Survey, January 25–28, 2002. Available at: www.russiavotes.org.

34. "Political News a Turnoff for Most Russians," and "Voters Trust Own Experience More than Media," *New Russia Barometer VIII*, January 13–29, 2000. Available at: www.russiavotes.org.

35. Bovt, "The Russian Press and Civil Society."

36. "Putin's Inaugural Address: 'We Believe in Our Strength,'" *New York Times*, May 8, 2000, p. A3.

37. "The President's Guideposts," *Trud*, July 20, 2001, pp. 1–2. From *CDPSP*, vol. 53 (August 15, 2001), p. 10.

38. ROMIR, "Sotsial´no-Politicheskaia Zhizn´ Rossii v Zerkale Obschestvennogo Mneniia," *Informatsionno-analiticheskii Ezhemesiachnii Biulleten´*, *2001 god.* Available at: www.romir.ru.

39. ROMIR, "Sotsial´no-Politicheskaia Zhizn´ Rossii v Zerkale Obschestvennogo Mneniia," *Informatsionno-analiticheskii Ezhemesiachnii Biulleten´*, *2001 god.* Available at: www.romir.ru.

40. "Russians Against Curtailing Freedom of Speech," VCIOM Survey, October 27–30, 2000. Available at: www.russiavotes.org.

41. Yelena Tokareva (interview with Yuri Levada), "All the Authorities Care About is Themselves," *Vremia MN*, July 13, 2001, p. 11. From *CDPSP*, vol. 53 (August 8, 2001), p. 6.

For Further Reading

Ernest Gellner, *Conditions of Liberty: Civil Society and Its Rivals* (New York: Penguin, 1994).

Nicolai Petro, *The Rebirth of Russian Democracy: An Interpretation of Political Culture* (Cambridge: Harvard University Press, 1995).

Harry Eckstein, Frederic J. Fleron Jr., Erik P. Hoffmann, and William M. Reisinger, *Can Democracy Take Root in Post-Soviet Russia: Explorations in State-Society Relations* (Lanham, MD: Rowman and Littlefield, 1998).

Timothy Colton, *Transitional Citizens: Voters and What Influences Them in New Russia* (Cambridge: Harvard University Press, 2000).

Nikolas Gvosdev, *Emperors and Elections: Reconciling the Orthodox Tradition with Modern Politics* (New York: Troitsa Books, 2000).

Marcia Weigle, *Russia's Liberal Project: State-Society Relations in the Transition from Communism* (University Park, PA: Pennsylvania State University Press, 2000).

David Hoffman, *The Oligarchs: Wealth, Power, and the New Russia* (New York: Public Affairs, 2002).

Christopher Marsh, *Russia at the Polls: Voters, Elections, and Democratization* (Washington, DC: CQ Press, 2002).

Christopher Marsh and Nikolas Gvosdev, eds. *Civil Society and the Search for Justice in Russia* (Lanham, MD: Lexington Books, 2002).

University of Strathclyde Centre for the Study of Public Policy and Russian Center for Public Opinion and Market Research, *Russia Votes* website, available at: www.russiavotes.org.

9

Russia's Struggle for Democracy

Joel C. Moses

Russia and Democracy

The greatest single overriding challenge that Russia has faced since the collapse of the Soviet Union in 1991 has been to institute political changes leading to a real democracy as a system of governance in the country. All of the security concerns in foreign policy and all of the domestic policy problems for Russia's leadership since 1991 in one way or another bear on this fundamental challenge of the country's uncertain democratic transformation.

Like eleven of the other fifteen post-Soviet countries, Russia never existed as a national state prior to 1991.[1] Therefore, Russia is not just like Chile, Korea, Taiwan, or Spain, countries that have successfully dismantled authoritarian regimes and instituted far-reaching political and economic changes. The political leadership of Russia, like those in the other new post-Soviet countries, has confronted the additional major task of nation-building, which must be achieved simultaneously with the other changes undertaken in their countries. They must, by their domestic and foreign policies, generate a sense of civic nationalism, if for no reason than to sustain the loyalty and support on the part of the populations residing under their authority to the very idea that they belong to a national community sharing a common fate and future. The essential issue for the Russian political leadership since 1991 has been and remains to justify and rationalize Russia's very existence as a country to its own people.

Russia is different in one important way from the other fourteen post-Soviet countries. The reason for the existence of an independent Latvia, Lithuania, Estonia, Georgia, Armenia, Azerbaijan, and the others was ethnic national self-determination. Their claims to sovereignty were responses to the accumulation of grievances by their majority ethnic nationality that led them to break from the Soviet Union in 1991. Their rationale as independent countries was to provide states for their ethnic nations. The same logic and rationale has never applied as much to the Russian Federation. With more than 25 million ethnic Russians residing outside of Russia in the other fourteen post-Soviet countries since 1991, Russia as a national state cannot justify itself as the homeland of ethnic Russians. Nor was ethnic self-determination ever the sole or even the major driving force for those like the Democratic Russia movement in 1990–91 and leaders like Boris Yeltsin who pushed for the sovereignty and greater independence of Russia from the Soviet Union. They were advocating not ethnic self-determination but a "democratic" Russia free from the authoritarian Communist system.

Russia's justification as a national state in this sense has always been closely connected to the ideal of democracy. The nature of Russia's political system as a democracy has been the reason for the country to exist. Presidents Yeltsin and Putin since 1991 have repeatedly said as much. They have contended that the real security and national interests of their country externally are to achieve conditions facilitating democratization and a viable market economy internally. Democratic conditions envision a country in which citizens share uniform rights and freedoms; political participation allows citizens a real effect through elections and other means to make their public officials accountable for their actions and truly responsive to majority will; and constitutionalism and a rule of law elevate the sense of shared community among all Russians based on tolerance, political pluralism, and civil liberties. A viable market economy ensures a wide range of opportunities for the highly educated Russian people in a country with seemingly unlimited natural resources as consumers, entrepreneurs, farmers, and workers to assume their role as a major world participant in a global economy of the twenty-first century. As President Putin told a national audience in 2001 (on the eleventh anniversary of the declaration of sovereignty passed by the Russian parliament): "This document marks the beginning of our new history. It is the history of a democratic state, based on civil liberties and supremacy of the law. The main goal of this state is the success, prosperity and happiness of its citizens."[2]

Explicit in Putin's remarks, just as they were a constant theme by his predecessor Boris Yeltsin, is that Russia's rightful claim to sovereignty as a country and ultimate national security depend on the achievement of these goals. To achieve the success, prosperity, and happiness of its citizens, Russia in turn must prove itself to be a trustworthy and reliable country. In the twenty-first century of globalization, trustworthy and reliable countries sought as economic partners and markets are governments widely viewed by others as stable and predictable just because they are democracies.

In foreign policy, Russian security and the effective promotion of its national interests ultimately depend on how other countries come to perceive the Russian government as a normal, civilized state. Other countries must come to view Russia as governed by the same rules of conduct as themselves, constrained by constitutional principles and a rule of law, and accountable to its own public for its actions. Only to the extent that Russia comes to be accepted by the West as a normal state can Russia even hope to solve the numerous economic and social problems for its public like widespread poverty, unpaid wages, an adequate food supply, deteriorating health care, increasing mortality, and declining population.[3] That acceptance alone will encourage foreign investment and trade in Russia and willing cooperation on the part of other countries—all so critical to Russia's very future. A democratic reality is also absolutely necessary to generate and maintain the public trust and confidence in their political institutions on the part of the Russian public in order for their government to be able to deal with these numerous problems over the long run.

The rationale by Russian leaders has been essentially the very same since 1991. The purpose of Russian foreign policy is not just to ensure the territorial security of the country by stemming the threat of ethnic nationalism along Russian's southern borders or by promoting Russian "great power" status and military-economic influence with the United States, other countries of the former Soviet Union, and East Asia.[4] These are only means to an end. The purpose is to create conditions through Russia's interactions in the international realm that positively contribute to the long-term goals of the country as a whole. Attainment of those long-term goals is essential to justify Russia as a national and political community to the Russian citizenry.

There has always been a consistent linkage between international security and national interests for Russia and the internal goals at least rhetorically advanced by Russian leaders since 1991. This is true despite differences between the ideological left, right, and center in Russian politics over the past decade and despite alternative approaches about how Russia can best achieve those goals, marking the pro-Western idealism of Russian foreign policy under President Yeltsin and Foreign Minister Kozyrev during 1992–95 from the more pragmatic realism focusing on the countries of the former Soviet Union by Yeltsin after 1995 and President Putin since 2000. Putin's support for the international alliance against international terrorism since September 11, 2001, coincidental with his aspiration to make Russia a member of the World Trade Organization, has only reaffirmed this essential linkage between Russia's foreign and domestic policy goals.

Thus, it is self-evident that one of the major reasons foreign investment in Russia has been so minuscule in per capita or even absolute terms compared to other post-Communist countries over the past decade is the lack of confidence on the part of the world financial and corporate community in the democratic transformation of Russia. In essence, Western financial and corporate interests have "voted" no in their assessment of Russia's internal transformation since 1991. The low level of Western financial investment has been a "vote of no-confidence" in the nature of Russian political institutions and the perceived absence of any assured real constitutional form of government and rule of law. Further, the lack of significant Western involvement in the Russian economy has been a "vote" against the lack of transparency, predictability, and other democratic attributes in Russia, which are examined in more detail in the sections below.

Imperfect Democracy

Russia, over the first decade of its existence, would be considered at best a very imperfect and flawed democracy. With the exception of some guarded optimism about the viability of civil society in Russia,[5] any assessment about the state of Russian democracy after the first decade would have to conclude on a balanced but overall pessimistic note. The forms of democratic institutions and processes have been established, but the norms and behavior of political leaders and the actual substance of daily political life in the country have fallen short of Western democratic ideals.[6]

On the one hand, the authoritarian institutions of the past Soviet Communist system have been permanently abolished, the subjugation of peoples and nations typical of the Soviet past can no longer be recreated, and the former Soviet nationalized command economy in Russia has been dismantled and privatized.[7] Russians are free to speak, practice their religions, vote in competitive elections, run for political offices, travel outside the country, and join groups and political parties. On the other hand, the goal of democratic consolidation still remains uncertain in Russia's future. The failure to achieve this consolidation threatens to nullify all of the positive changes that have occurred in the country. Russia's national security—defined in terms of democratization and economic integration in the world global economy and universally proclaimed as their ultimate goals by Russian leaders since 1991— remains at risk. The sections below will survey some of the shortcomings in Russia's democratic transition and the challenges remaining to overcome.

Elections and Public Offices

Nothing more clearly illustrates the difference between democratic form and undemocratic reality in the Russian context since 1991 than elections and elected public offices. On the positive side, competitive elections with secret

ballots to choose those who would govern them have now become an un-
alienable political right. In contrast to the past Soviet era, the Russian public
who still vote in large numbers even in local municipal and regional elections have come to expect a right to determine both their leaders and policies. Elected representatives to local and national assemblies and elected
chief executives as their mayors, governors, and presidents derive their ultimate formal authority to govern from the Russian public, and the legitimacy
of Russian political institutions rests on the majority will, formally expressed
through the ballot-box as the consent of the governed.

In certain ways, Russian voters are become both rational and savvy, able
to make distinctions on the merits of candidates and political parties relative
to their own preferences for policies and priorities. Similar to voters in Western democracies, Russians are motivated in their voting more by whom and
what they oppose than whom and what they support. At a minimum, voters,
when they see little real choice among candidates, have even exercised their
option of invalidating the election entirely to show their displeasure. The
Russian electorate can do so by not showing up to vote in large enough numbers so that the total percentage of registered voters participating falls below
the 50 percent legal minimum requirement to validate an election; or they
can mark their ballots on the line in all elections termed "Against All (*protiv
vsekh*)," which, if a majority of those voting do so, also nullifies the election.[8] At a maximum, voters given a real choice over the candidates for public offices do show up in large numbers and have ousted even very corrupt
and autocratic chief executives like governors, unswayed by the large amounts
of money poured into the campaigns or the manipulation of the media and
their offices by incumbents.[9] A minority of incumbents for Russian governor
lost their reelection bids in 1999–2001, but the very fact that some incumbents lost or—anticipating losing—withdrew does indicate the power of the
Russian voter as a very positive democratic change since 1991.

On the negative side, the ideal of elections as an instrument of public
empowerment in a democracy differs from the more sordid reality surrounding many elections in Russia. Elections in Russia are free, but they are far
from fair. Elections almost as a norm in Russia have become marred by scan-
dal, dirty tricks, fraudulent practices, illegalities, and outright manipulation.[10]
Campaigns for public office in Russia have almost taken on the form of open
"guerilla warfare" between candidates, in which the only rules have become
defeating your opponents by any means possible.

Candidates distribute or encourage through friendly media false or com-
promising stories about their opponents (*kompromat*), or themselves publish
phony editions of newspapers with headline stories discrediting their rivals
(*falshivki*). Professional consultants are hired for exorbitant amounts of money

to wage a campaign of dirty tricks (*chornaia tekhnologiia*) and negative advertising (*chornyi-PR*) against their opponents. Vote-tampering and bribing of local election commission members and judges to remove candidates from the ballot are commonplace. Illegal campaign funding by local oligarchs or crime bosses is generally assumed. Journalists are paid to write or air stories favorable or unfavorable about the candidates (*pechatnaia zakazukha*). Potential voters are enticed with the promise of inexpensive consumer goods sold near precincts the day of elections or actually paid to mark their ballots for candidates. Campaign billboards, leaflets, and posters are destroyed by the opposing sides. Campaign headquarters of rival candidates are targeted for mysterious fires or bombings. And campaign workers or even candidates themselves are beaten and sometimes even murdered before the election.

Fraudulent elections have become so much the norm that in the many recent gubernatorial elections held throughout Russia in 1999–2001, the Russian media labeled similar forms of campaign dirty tricks and illegalities perpetrated by candidates to fix the outcome of elections in many parts of the country. The labels referring to these patterns of dirty tricks and illegalities to fix elections have now become part of the Russian political lexicon.[11] The most that can be said is that the "democratic bar" of how campaigns and elections should be conducted fairly is being raised over time in Russia. The reason is that there are so many reports about these irregularities in the Russian media and so many candidates sue each other for violating campaign rules and regulations before their local electoral commissions.

The nominal right for all Russians to run for political office also remains something more of an abstract promise than a reality. The political playing field is far from equal for all. Campaigns for political office even at the regional level now cost the equivalent of hundreds of thousands of dollars. Political parties are still widely discredited in the minds of most Russian voters, who rank them in public opinion surveys as the very most distrusted institution in Russian society.[12] Therefore, average citizens with little personal wealth who gain the endorsement of a political party only through party affiliation are less likely to win. Those positioned to run and win political offices in Russia must already have accumulated a great deal of personal wealth on their own or at a minimum have ties to others with the money to underwrite their campaign. They run as candidates unaffiliated with any political party and appeal to voters more in terms of their personalities than their programs or policy platforms.

As a consequence, the majority of individuals now running for offices as mayors, governors, or deputies to assemblies are owners of businesses, many of whom took advantage of the privatization of Russia's economy after 1991 to make their fortunes often unscrupulously in the auctions selling off formerly state-owned sectors. They have come to monopolize local economic sectors, like supermarkets and banks, and can take advantage of their high name-recognition among the public, reputation deserved or not as a successful manager, ownership of local newspapers and televisions stations, and personal wealth to mount their usually successful campaigns. In some cases, their electoral success benefits from their close links to organized criminal syndicates in their locales. The criminal syndicates help underwrite their campaigns, deal violently with their opponents, and expect favors in return from the same public officials whom they have helped put in office.

Governance

Despite the promise of enhancing local democracy by giving the Russian public since 1995 the right to directly elect their mayors and governors, the potential has failed to materialize in its fullest sense. The process to recruit chief executives has changed from the prior Soviet Communist era. The mindset and behavior of the chief executives as mayors and governors have not. Autocratic leaders, who do not differ significantly from the Communist Party *apparatchiki* of the past, have become more the rule than the exception in many locales throughout Russia. Elected officials display little sense of public virtue or any moral compunction about public service other than using their elected offices to enrich themselves and their political friends and promote their own interests first and foremost.

A collaborative study of the Yaroslavl oblast was conducted by Western and Russian social scientists over a decade.[13] They found a personalistic style of rule by chief executives who are contemptuous of institutions, rules, and procedures.[14] It is a style of governance all too similar to that practiced for decades by the Communist *apparatchiki* in the previous Soviet era. The only difference from the past Soviet era is that chief executives today, unlike former Communist officials, now have to win the authority to act in this way by first prevailing over their opponents in public elections.

There is very little political accountability of the dominant chief executives once elected to office. The political reality in many Russian locales like Yaroslavl is little more than a "delegative democracy," in which officials do pretty much what they want after they have won the formal endorsement to govern through elections by the public.[15] Power is concentrated in the offices of the mayors or governors (or presidents as the chief executives are termed in Russian ethnic republics). Mayors or governors are seldom checked by local media, which are financially subsidized or even directly owned by themselves. They have little fear of judges in regional courts who are underpaid and dependent on departments administered by the mayors or governors for their own housing and other material benefits. Most importantly, mayors and governors in many locales can presume that the majority of deputies in their legislative assemblies will be compliant to their will as the chief executives.

Regional legislative assemblies are nominally an independent branch of government. They are supposed to exercise legislative oversight over the conduct and actions of the executive branch, initiate and pass legislation responsive to the interests of their constituents, and negotiate the terms of the budget with the mayor or governor. In reality, many Russian legislative assemblies are dominated by members of the regional economic-business elites or by heads of districts politically dependent on the mayors or governors for their positions. They are all allied in promoting and protecting their collective establishment interests on the regional assemblies in concert with the chief executives and indifferent to the consequences of their actions for the public at large.

Civic Culture and Political Pluralism

Opinion surveys analyzed over a decade in Yaroslavl demonstrate that deteriorating economic conditions since 1991, along with the disillusionment over their elected officials, have eroded much of the Russian public's early enthusiasm and support for the new democratic institutions.[16] The findings in Yaroslavl coincide with similar conclusions from national samples of Russians during the same period.[17] The public is increasingly distrustful about their political leaders, even though they retain a fairly high level of support for the political institutions, the formal democratic processes, and at least abstractly for most democratic values. The Russian public still appears to have high democratic expectations and beliefs, but their growing disillusionment with democratic institutions has more to do with the failure of their elected leaders to live up to those expectations.[18]

Subsequently, there has been a clear erosion in the general attitudes political scientists would equate with a democratic "civic culture" in Russia over the past decade. Russian's willingness has fallen off sharply to join organizations and participate politically in ways other than voting, in stark contrast to the millions who joined groups, protested, and otherwise demonstrated their sense of political efficacy and self-confidence in the late 1980s and early 1990s.[19] The Russian public's "withdrawal from politics," other than by voting, reflects a return to an attitudinal division between the political elite and average Russians ("dual Russia") similar to Russia's tsarist and Soviet past, when citizens meekly and unquestioningly deferred to the will of authorities even though they distrusted and despised the same officials.

The very word "democracy" itself has become almost a term of derision for many Russians. The word is negatively associated in their minds with the capitalist market reforms and widely despised economic liberal reformers like Anatolii Chubais and Yegor Gaidar. Their policies to free prices and privatize the economy are blamed for the widespread poverty, crime, and other societal ills afflicting the country since 1991. The only politicians and political parties in Russia still openly willing even to label themselves "democrats" are the Union of Rightist Forces and Yabloko. Yet, in the most recent election to the State Duma in December 1999, these two political parties combined were only able to garner 15 percent of the national party-list vote (Russians choose one-half of the Duma's 450 deputies by a proportional-representational ballot, and the other one-half through single mandate elections). Those same political parties and the very few active democratic mass movements throughout Russia remain, even after a decade, disorganized, weak, and torn by internal squabbling among their leaders.[20] Political pluralism as an essential feature of democracy has come to mean in Russia competition among rival economic oligarchs and financial institutions. They sponsor and pay for candidates to run for political office to ensure special tax benefits, lucrative contracts, and other benefits for themselves.

There are pockets of political pluralism and more lively citizen political activism and participation in Russia. The two most prominent and conspicuous are the cities of Moscow and St. Petersburg. There, one finds highly cosmopolitan civic political cultures with many active political parties, interest groups, and involved citizenry. However, this reality is unrepresentative of Russia. The civic political culture in Moscow and St. Petersburg, results from their large populations, their roles as the national and cultural capitals of the country, a large percentage of highly educated professionals, beneficiaries from most of the Western investment in Russia since 1991, and headquarters for the most powerful financial-corporate institutions in the country. As the democratic showcases of Russia, Moscow and St. Petersburg, have also undergone significant changes to become physically more Western in their appearance over the past decade, evidenced by the construction of shopping centers, supermarkets, high-rise office buildings in Moscow, and the renovation of many pre-1917 buildings in St. Petersburg to attract foreign investment and tourism.

Yet, even for Moscow and St. Petersburg, there is another side to their seeming Westernization in politics and physical appearance. Moscow has been governed for the past decade by its mayor, Yuri Luzhkov, and St. Petersburg since 1996 by its mayor, Yegor Yakovlev. Both Luzhkov and Yakovlev have become nationally notorious for their autocratic style of governance, their promotion of dubious and highly expensive urban development projects that financially benefit contractors and others politically allied with them, and, especially for Yakovlev, murky ties to organized criminal syndicates.

Civil Liberties

In Russia, civil liberties and a free press also remain fragile uncertainties. The reasons are a prevailing Russian-public willingness to defer to political authorities, the widespread use of law and courts as political weapons by public officials against their opponents, and a warped sense of public ethics condoning the bribing of officials and journalists. A cowed independent media has withdrawn into a form of self-censorship in the level of its criticisms and oversight of public officials. Self-censorship has become a necessity to protect the owners and editors of the few still remaining independent media against economic pressure through hostile stockholder takeovers of their corporations, libel suits to bankrupt their operations, or other forms of legal manipulation to close them. The economic pressures are often entirely orchestrated behind the scenes by the same public officials.[21] Further, Russian public officials, business-owners, and journalists have all become frequent targets of contract murders. Very rarely, if ever, will the perpetrators of these crimes be discovered, let alone arrested and tried. Thus, civil liberties for Russians to speak and act freely are inhibited by an environment in which those with whom you disagree can murder with relative impunity.

The one positive democratic counterpoint to the decline in civil liberties for Russia has been the spread of the Internet. Although probably at most one-tenth of the Russian population has access to computers and the Internet, the younger generation of Russians has become wired through computers more readily available at schools, libraries, and "Internet clubs," which exist in most medium-size and large cities. The growing importance of the Internet is evidenced by the fact that in mid-2001 all Russian ministries were advised that they should update the information available at websites weekly. Access to the worldwide media through the Internet for these Russians has reduced the ability of the Russian government to intimidate the remaining independent Russian media and to impose a form of self-censorship. In addition, newspapers in several Russian regions post their stories and editorials on the Internet. By positing them on the Internet, the newspapers can get around the financial pressures and other attempts to censor them by local government officials. More Russians have come to depend on the Internet rather than on the costly hard copies of print media for their sources of information.[22] Not coincidentally, the Union of Rightist Forces and Yabloko have sponsored the increasing availability of computers and the Internet in schools and libraries throughout Russia as major goals to publicize their political parties. They have been supported in these efforts by grants from the Soros Foundation and other international philanthropic organizations committed to aiding democratization in Russia.

Organized Crime and Police State

The widespread acceptance of elected public officials as the basis of government legitimacy for the Russian public has been discredited and undermined by a political reality in which many of the same elected officials have been corrupted and intimidated by powerful organized criminal structures throughout Russian society.[23] Russian democracy has been eroded in a society where criminal syndicates have come to dominate, penetrate, and capture almost all spheres of political and economic life in the country. Power rests not with the publicly elected institutions in Russia but with the informal structures of organized criminal syndicates.

More importantly, one could argue that organized criminal syndicates truly reign in Russia. The criminal syndicates are seemingly immune from prosecution and accepted as a way of life by both the public and the political elite. Rather than the democratization of Russia over the first decade, we have witnessed the criminalization of Russian society. Rather than the advancement of a rule of law, we have seen its perversion into a rule literally of the mob—that is, the mafias that informally have divided up territories and criminal enterprises among themselves, resolve conflicts among themselves through contract murders, and maintain a high level of fear and intimidation over political and economic leaders behind the scenes. Russia has been informally transformed into something equivalent to Colombia—a gangster state in which the nominal democratic structures have been essentially captured by organized criminal mobs. Indeed, given the increasing cooperation between Russian and other, international criminal gangs inside Russia, which take advantage of the country's corrupted law-enforcement system to orchestrate their worldwide criminal activities without fear of arrest, Russia as the international capital of organized crime is becoming a major international security threat to the world.

An underlying police-state mentality has reemerged over recent years in Russia, coincidental with the rise to power of the former KGB officer Putin as Russian president and the numerous former KGB officials whom he has promoted to high political offices throughout the country. The unmasking of spies and other domestic enemies of the state, now labeled as terrorists by security and military institutions, has become a more prevalent motif of daily political discourse in Russia. Anyone who has spent any time in Russia over the past few years could not but have noticed the change toward a more police-state atmosphere. The police are a much more visible presence on the streets in Russian cities. Packages of commuters on trains and subways are constantly being checked by the security guards. And people with dark skin are frequently stopped, frisked, and asked to show their internal passports.

The mentality of a besieged state has been used to justify taking extreme judicial measures against individual Russian citizens in order to protect the greater Russian public. Environmentalists and journalists like Aleksandr Nikitin and Grigorii Pasko, who have disclosed and reported on the dumping of nuclear wastes by the Russian military, or Russian social scientists like Igor Sutiagin, who collaborate with Western research institutes on projects deemed "sensitive" by the Russian military or security organs, were arrested. They were tried and imprisoned for the unlawful disclosure of state secrets and even treason in a chilling return to an atmosphere associated with the Soviet past. For these individuals to be singled out and tried is almost the equivalent to the show-trials of a past Soviet Communist era and sets a dangerous precedent. The actions against Nikitin, Pasko, and Sutyagin also raise additional doubts about the sincerity of Russian political leaders since 1991 to institute a Western legal system that guarantees all Russians due-process rights and equal treatment under the law in court.

The increasing pressures by the Russian national government to impose self-censorship on the media, and the heightened sense of security against internal enemies and spies, largely stem from the Russian military's war against Chechen separatists in Chechnya since 1999. The justification for the military action has been to eliminate a Chechen government perceived in Moscow as promoting terrorism and Islamic fundamentalism. This second Chechen war (the first was during 1994–1996) has accounted for thousands of Russian military casualties since 1999. It has also resulted in brutal atrocities and widespread human rights abuses committed by the Russian military against the Chechen civilian population. The Russian military action itself was a direct response to a series of so far unsolved bombings of Russian apartment buildings in September of 1999. The Russian government blamed Chechen terrorists for the bombings. Others, including Boris Berezovsky and former security officers, have claimed that the bombings were actually committed by the Russian security forces to provide a provocation and pretext for the Russian military action then launched in Chechnya.

Whoever is to blame for the bombings, and whatever the justification for the Russian military action in Chechnya, the war in Chechnya directly accounts for the heightened sense of insecurity, the mania about terrorists and spies, and the general return to a police-state atmosphere threatening democracy overall in Russia. Particular targets of the pressure orchestrated by the government have been those independent national newspapers and television networks which in 2000 and 2001 still attempted to report objectively on the war in Chechnya, including the widespread abuses against the civilian population by the Russian military forces stationed there. The political cover for the government was that the same media were owned by wealthy billionaires (Vladimir Gusinsky and Boris Berezovsky). The Russian public had come to view both Gusinsky and Berezovsky very negatively, as odious figures who had accumulated their vast fortunes illegally in the 1990s through their personal ties to President Yeltsin. Thus, the shutting down of their independent media, as well as the self-imposed exile of both Gusinsky and Berezovsky, has actually been supported by a majority of the Russian public as justice meted out against corrupt billionaire oligarchs rather than an attack on the free press.

Who Is to Blame?

If the overall state of democracy were to be assigned a letter grade since 1991, Russia would probably receive something on the border between a C– and a C. In conventional Russian and Soviet political culture, whenever problems or mistakes would arise in carrying out the goals set down by the government, the normal first response would be not to find solutions but to assign fault. The question asked when things went wrong was not what should be done but "who is to blame" (*kto vinovat*).

Were Russians to apply the same logic of assigning blame to explain their marginal democratic progress since 1991, it would start at the top of the political system. The nature of democratic governance in a country extends downward from the institutional framework established at the national level and from the behavior of those whose national offices are entrusted with primary authority. Both the institutions and the conduct of the most prominent national political leaders set a model of governance for others throughout the country to emulate in how they exercise power and in how they act in their offices. A democracy must strike a fine balance. Institutionally, a democracy must have a strong national government, able to carry out the will of the public. But at the same time a democracy must also be a system in which those entrusted with so much authority act constitutionally and accept the constraints and checks on their power, and restrain their own human tendency to aggrandize power in their own hands.[24] Under the 1993 Russian Constitution, the Russian national government is a super-presidential system, under which the office of president is granted an excessive amount of authority with very weak affective countervailing constraints and checks on the office by the legislative branch of government.[25] The executive-dominant system seen throughout Russia since 1991 by mayors or governors over their weak local assemblies directly reflects the same super-presidential nature of government at the national level established under the 1993 Russian Constitution.

The president has enormous powers vis-à-vis the main legislative body, the State Duma, the head of government, the prime minister, political parties, or the courts. Elected to a four-year term, the president has the constitutional authority to appoint the prime minister from the State Duma, regardless of the distribution of seats held by various political parties from elections to the State Duma. The president can dissolve the State Duma if it rejects his nominee for prime minister three times. If the Duma passes a no-confidence vote in the policies of the prime minister, the president can dissolve the Duma. Moreover, the prime minister is not an independent actor, but rather works directly for the president and can be fired by the president at any time. Further, the president appoints and removes the most powerful cabinet offices under the prime minister (the ministers for foreign policy, defense, the security police, and the national civilian police). Political parties in Russia are weak, beset with internal squabbling and disagreements. This situation allowed President Yeltsin to get his way over most legislation introduced by his administration during the nine years of his presidency. The executive branch under Yeltsin was also not accountable to the courts. By these and other constitutional powers vested in the office of Russian president, such as the right to issue presidential decrees, the executive branch is very strong, but with few equivalent and real countervailing checks and constraints on the presidency itself.

For Boris Yeltsin, the dominant powers vested in the office of Russian president proved to be addictive. The office of president for Yeltsin unleashed the very worst traits of his personality and encouraged him to govern in a capricious and autocratic manner, influenced more by his wealthy oligarch friends and political cronies than by constitutional rules and procedures. Yeltsin's style of governance from 1991 through 1999 has been characterized as the reign of an autocratic and almost tsarist-like "patriarch" rather than a democratic president.[26] Yeltsin in turn set the tone for how other chief executives like governors, republic presidents, and mayors perceived it correct to act throughout the 1990s.

Through treaties with republics and regions and numerous arbitrary grants of special economic concessions or favors to different locales, Yeltsin contributed to a political climate in which chief executives throughout Russia were encouraged to get as much autonomy from the center as they could and to govern just as arbitrarily and indifferently to any rules of democratic conduct and law as Yeltsin was dealing with them. Yeltsin's decision to sign a treaty and grant certain kinds of allowances to a particular Russian locale was calculated solely by his own shifting political priorities. Yeltsin violated the very equality presumed in the Russian Constitution of 1993 for all eighty-nine administrative subdivisions by singling out certain locales over others for preferential treatment in treaties, and intentionally or not, he sent a message to other chief executives throughout Russia. The message was that they should feel only minimal formal obligation to comply with the constitution and should conceive of their primary responsibility to their own local power elite rather than to the Russian public or Russian national state.

Since 1999, President Putin has done two things. First, he has further strengthened the Russian presidency institutionally. President Putin has benefited from very high public approval ratings, which makes it difficult for any politician in Russia to question his policies or to mount opposition against them. President Putin also governs with an even more docile parliament than the Yeltsin era. The docility of the Russian parliament has arisen partly because a pro-Putin party won the second largest number of seats in the December 1999 State Duma election and partly because Putin pushed through the parliament in 2001 a reform in selecting members for the upper chamber (Federation Council), weakening its ability to oppose presidential initiatives.

Second, since 1999 President Putin has attempted to correct the worst abuses of "Yeltsinism" as a style of governance by various reforms to reassert more centralized political authority and to rein in the autocratic and corrupt chief executives in the administrative subdivisions throughout Russia.[27] For Putin, Russia politically under Yeltsin had begun to unravel. Putin has defended his reforms to institute a stronger national state by arguing that they are intended primarily to advance real democratic equality, liberties, and rights for all Russian citizens.[28]

By Putin's youth, activism, and pragmatism, his style of governance is in every way different from Boris Yeltsin as president. Yet, like Yeltsin whose attitudes were formed from his past as a Communist Party official, Putin is a former career KGB officer whose true internal commitment to democracy in Russia still remains questionable. However sincere Putin may be in equating a strong Russian national state with democratic freedoms for all Russians, Putin is also someone who since 1999 has overseen a shift in the political atmosphere of Russia outlined earlier in this chapter. The Russian national government since 1999 under President Putin has become less tolerant of a free independent press and has reduced the civil liberties of Russians.

In contrast, whatever might be said about ex-President Yeltsin, his presidency contributed positively to an era in which Russian society seemed to be freer. Yeltsin and all politicians came to accept being criticized for their actions and policies in the Russian media, being investigated in parliament for their misdeeds, being challenged in tell-all books by their former political advisers, and even being lampooned weekly on Russian national television satire shows. Political pluralism seemed much more a reality under President Yeltsin in 1991–99 than it has been since 2000 under President Putin. For Putin, who allegedly has a large portrait of Peter the Great adorning his office in the Kremlin, his vision of Russia seems to be more as a great and powerful state than as a political democracy.

Thus, democracy is still very much a work in progress in Russia. Elections and elected public offices represent the irrefutably positive changes and inalienable political rights attained by all Russians since 1991. The real indeterminants are how far these incubators of democracy are allowed to reproduce the other prerequisites of a real enduring democratic form of governance in Russian society. How far Russia advances democratically still depends ultimately on the powerful office of the Russian president and the particular personality of the individual entrusted with the authority of that office.

A "managed democracy," which appears to be the goal of President Putin for Russia, is an oxymoron. That is the essential problem of democracy in Russia. Democracies cannot be instituted by reforms from above by a leader. Civil societies cannot be ordered by presidential decree. The political accountability and responsiveness of elected officials to the public cannot be dictated. And tolerance and political equality cannot be willed by presidential edict.

Notes

1. The exceptions are the three post-Soviet Baltic states of Lithuania, Latvia, and Estonia, independent from 1919 until 1940. Russia existed under the Russian Empire until 1917, not as a distinct sovereign country and independent national state in the modern sense, but as the largest core of the Russian Empire, which geographically and politically ruled over the territories and populations constituting the other fourteen post-Soviet independent countries, Finland, and eastern Poland.

2. "Russian President Says Russian Democracy Here to Stay," Russian Public TV (ORT), BBC Monitoring, June 12, 2001, reproduced in Johnson's Russia List, no. 5296, June 13, 2001. Available at www.cdi.org/russia/johnson.

3. See the chapters on labor challenges (by Debra Javeline), rural revival (by Stephen Wegren, Vladimir Belen'kiy, and Valeri Patsiorkovski), and on demographic problems (by Timothy Heleniak) in this text.

4. See the chapters by Gregory Gleason, John Reppert, Mikhail Alexseev, and Dale Herspring in this text.

5. See the chapter by Christopher Marsh in this text.

6. For a balanced perspective of Russia's democratic political changes since 1991, see Archie Brown, "Evaluating Russia's Democratization," in Archie Brown, ed., *Contemporary Russian Politics: A Reader* (Oxford, UK: Oxford University Press, 2001), pp. 546–68.

7. Michael McFaul, "Two Out of Three Is Not Good Enough," *Moscow Times*, reproduced in Johnson's Russia List, no. 6044, January 28, 2001. Available at www.cdi.org/russia/johnson.

8. As one well-documented example, voters in the million-plus Russian city of Novosibirsk boycotted the city council election in large enough numbers to invalidate the elections to nineteen of the twenty-five city council seats on December 10, 2000, as a sign of their displeasure against the same city council raising the fares for public transportation within two weeks prior to the election. When new elections in the nineteen districts had to be called in April of 2001, fares became a major issue during the campaign by candidates in response to the December voter boycott.

9. Joel C. Moses, "Political-Economic Elites and Russian Regional Elections, 1999–2000: Democratic Tendencies in Kaliningrad, Perm, and Volgograd," *Europe-Asia Studies*, vol. 54, no. 6 (2002), pp. 905–31.

10. Ibid.

11. One such dirty trick by candidates has been arranging for individuals to be placed on the same ballot with the exact same name as the individual whom they want to defeat (termed *dvoiniki* or doubles in Russian) in order to confuse voters into marking the wrong name in the voting booth.

12. At the beginning of 2002, only 10 percent of Russians in a national poll trusted political parties, while 75 percent distrusted them: Nikolai Popov, "Trust but Verify. About Confidence in Popularity Ratings and Trust in Them," *Novoye Vremia*, January 27, 2002, reproduced in Johnson's Russia List, no. 6045, January 28, 2002. Available at www.cdi.org/russia/johnson.

13. Jeffrey W. Hahn, ed., *Regional Russia in Transition: Studies from Yaroslavl'* (Baltimore, MD, and Washington, DC: The Johns Hopkins University Press and Woodrow Wilson Center Press, 2001).

14. Ibid., pp. 234–42.

15. Ibid.; and Andrei Tsygankov, "Manifestations of Delegative Democracy in Russian Local Politics: What Does It Mean for the Future of Russia?" *Communist and Post-Communist Studies*, vol. 31, no. 4 (October–December 1998), pp. 329–44.

16. Hahn, *Regional Russia in Transition*, pp. 75–137.

17. Thomas F. Remington, *Politics in Russia*, 2nd edition (New York: Longman Publishers, 2002), pp. 84–87, 123–28.

18. Timothy J. Colton and Michael McFaul, "Are Russians Undemocratic?" Excerpts from *Working Paper No. 20*, Carnegie Endowment for International Peace (June 2001), reproduced in Johnson's Russia List, no. 5294, June 12, 2001. Available at www.cdi.org/russia/johnson.

19. Remington, *Politics in Russia*, pp. 83–87.

20. An excellent study of one democratic mass movement that should have been able to generate widespread public participation and influence in Russia, but has not, is Valerie Sperling, *Organizing Women in Contemporary Russia: Engendering Transition* (Cambridge, UK: Cambridge University Press, 1999).

21. See Laura Belin, "Political Bias and Self-Censorship in the Russian Media," in Brown, *Contemporary Russian Politics*, pp. 323–42.

22. On the growth of the Internet in Russia, see the series of articles "Internet in the Regions," East-West Institute, *Russian Regional Report*, vol. 5, no. 3 (January 26, 2000). Available at www.iews.org/rrrabout.nsf.

23. See the chapter by Louise Shelley in this text.

24. On the Western democratic model of this balance between a strong government and constitutional checks and constraints on those who govern, see the chapter by Christopher Marsh in this text.

25. See Remington, *Politics in Russia*, pp. 51–68.

26. See George W. Breslauer, "Boris Yeltsin as Patriarch," in Brown, *Contemporary Russian Politics*, pp. 70–81.

27. See Eugene Huskey, "Overcoming the Yeltsin Legacy: Vladimir Putin and Russian Political Reform," in Brown, *Contemporary Russian Politics*, pp. 82–96; and the chapter by Darrell Slider in this text.

28. Vladimir Putin, "Putin Says No Threat of Police State," *Izvestia*, July 14, 2000, reproduced in Johnson's Russia List, no. 4407, July 16, 2000. Available at www.cdi.org/russia/johnson; "Internet Press Conference with Vladimir Putin," Johnson's Russia List, no. 5135, March 7, 2001; and Vladimir Putin, "Press Conference," Johnson's Russia List, no. 5354, July 18, 2001.

For Further Reading

Harry Eckstein, Frederic J. Fleron Jr., Erik P. Hoffmann, and William M. Reisinger, *Can Democracy Take Root in Post-Soviet Russia?: Explorations in State-Society Relations* (Lanham, MD: Rowman & Littlefield Publishers, Inc., 1998).

Valerie Sperling, *Organizing Women in Contemporary Russia: Engendering Transition* (Cambridge, UK: Cambridge University Press, 1999).

Archie Brown, ed., *Contemporary Russian Politics: A Reader* (Oxford, UK: Oxford University Press, 2001).

Jeffrey W. Hahn, ed., *Regional Russia in Transition: Studies from Yaroslavl* (Baltimore, MD, and Washington, DC: The Johns Hopkins University Press and Woodrow Wilson Center Press, 2001).

Thomas F. Remington, *Politics in Russia*, 2nd edition (New York: Longman Publishers, 2002).

PART THREE

SOCIOECONOMIC
CHALLENGES

10

Labor Challenges and the Problem of Quiescence

Debra Javeline

With the collapse of the Soviet Union came the impoverishment of much of the Russian labor force. Some lost jobs, and many more suffered hidden unemployment in the form of unpaid wages. Russians and observers of Russian politics increasingly predicted that the desperate plight of workers would ultimately lead to massive labor unrest and threaten the stability of the Russian state. They were wrong. While some strikes and protests did occur in Russia, especially in response to the so-called wage arrears or nonpayments crisis, the Russian labor force did not produce an organized, sustained effort to redress its grievances politically, and it certainly did not engage in riots and large-scale violent protests to challenge the state.

If the only challenge that workers could pose to Russian policy was social unrest, the story would end here. Russians did not protest widely and intensely in the late 1990s at the peak of the wage arrears crisis, so there is little reason to expect that they will protest now at the beginning of the twenty-first century when revenue from increasing world oil prices has seemingly revitalized the Russian economy. This revenue, while not rectifying the structural problems that created the wage arrears crisis, has alleviated the problem by enabling the paying off of some back wages and encouraging regular wage payment, at least in the short term.

However, social unrest is not the only challenge that workers could present for Russian policy in the twenty-first century. The quiescence of the labor force may seem desirable to political leaders trying to push through policy with minimal public resistance, but this very quiescence can also be detrimental in other policy realms. Specifically, the challenge that labor poses to Russian policy is in the social sphere. A workforce that does not try to redress its grievances politically seems to turn instead to physically and socially unhealthy behavior, and the effects are seen in increased rates of alcoholism, suicide, malnutrition, depression, and mortality. To the extent that a healthy, energetic labor force is needed to sustain a country economically and militarily, Russian policy makers cannot afford to be complacent, let alone satisfied, about labor passivity.

It is the aim of this chapter to examine the policy implications of a quiescent Russian labor force. First I will describe the situation of Russian workers since the fall of the Soviet Union. Then I will describe the origins of worker problems. These origins are highly complex, and I will argue that due to the complexity, most Russians have not attributed blame for their situation to a specific culprit. In the absence of blame or the presence of too much blame to go around, Russians have not protested. They also have felt a loss of control over the most important aspect of their lives, basic survival and well-being. This perceived loss of control or perceived helplessness on such a large scale can spell disaster for public health. Social psychologists have shown that the effects of helplessness on individuals include depression and other maladaptive behavior. I will argue that this relationship holds on a mass level in Russia and that policy must address the root problems of unemployment, unpaid wages, and poverty in order to remedy the resulting problems in public health and social welfare. Policy must also address the social psychological problems head-on and allow workers to take back some control over their destinies.

Russian Labor Problems

Many problems facing contemporary Russian workers are the problems of workers worldwide, such as unemployment, low wages, and inadequate benefits, and some problems, like the chronic delay or nonpayment of wages, are relatively unique to post-communist transition. The extent of these problems is hotly contested thanks to the notorious unavailability and unreliability of data.

Unemployment

One of the most hotly contested questions is the number of unemployed in Russia. Official sources report unemployment based on the number of individuals who officially register with the Federal Employment Service. However, most nonworking Russians do not register with the Employment Service, either due to shame or the perceived futility of using the service to search for another job, so true unemployment is far higher than the registration statistics reveal. In 1996, specialists on the Russian labor market determined that unemployment in Russia was close to nine percent of the population.[1]

Even if nine percent is a reasonable approximation of the actual unemployment figure, it still captures only part of Russia's employment problem. The very definition of "employed" is debatable in contemporary Russia, where workers frequently work without pay, work for pay below the subsistence minimum, get temporarily laid off, earn money through irregular casual work or self-employment, and engage in subsistence agriculture. A large percentage of the population is thus best characterized as in limbo between employment and unemployment. The number in limbo could be as high as 14 million people, or 19 percent of the economically active population.[2]

Unemployment rates vary tremendously by region and even between towns and cities within a single region, and in certain Russian regions, unemployment has been devastating. Hardest hit have been one-company towns that produced military-industrial goods that are no longer in demand or textiles that have lost in the competition with cheaper imports.[3] Women and men are equally likely to be unemployed in Russia, largely because women will remain in jobs with the lowest wages and worst working conditions.[4] A large percentage of the unemployed are young or of pre-retirement or pension age. Youth unemployment is explained by "the growing tendency for urban teenagers to live on their wits rather than to look for a job" or by turning to crime. Older Russians have generally been laid off with no chance of landing another job. There is in Russia "a stagnant pool of the unemployable."[5]

The unemployment problem is not alleviated by the activities of the Employment Service. Most unemployed Russians tend to have relatively high qualifications and skills, making them unsuitable for the few vacancies available that are mostly for unskilled jobs.[6] Remedying unemployment has been a low priority for the Russian government, probably because the actual state of affairs is concealed either deliberately or by the confused definition of unemployment.[7]

Nor is unemployment alleviated by vast employment opportunities in the new private sector. Most of these opportunities are available only to residents of Russia's largest cities, and even these are increasingly professionalized and competitive. Moreover, many private sector jobs are casual, providing no job security or social protection, so that it is unclear whether the job-taker really enters the ranks of the employed or merely moves from unemployment to the limbo between employment and unemployment.[8]

Low Wages

Those who are fortunate enough to be employed in Russia still may find it difficult to escape poverty. Wages can sometimes be so low that they fail to support even one person at the minimum subsistence level. A study of household income in November of 1996 revealed that 64 percent of households had a total income per head below the physiological subsistence minimum. In mid-1997, real wages had reached only 43 percent of the 1990 level.[9] Economist and former advisor to the Russian government Anders Aslund has declared that such low earnings, combined with falling productivity, present a problem for Russia far greater than unemployment.[10]

Wage disparities are most apparent between different Russian regions and between different branches of production. Rural residents are among the most poorly paid, with the average wage for this 27 percent of the population being far below the subsistence minimum.[11] Women are also among the most poorly paid, and they tend to work in the most depressed sectors like education and health care.[12]

The problem of low wages has not been alleviated by widespread engagement in secondary employment or casual labor. Those who have access to secondary employment tend to be better paid in their primary jobs, and casual labor tends to be the recourse of the unemployed more than the poorly paid. Casual labor, self-employment, and subsistence production may help avoid starvation but seldom promise a much higher standard of living.[13]

Wage Arrears

Even more vexing than the problem of low wages is the problem of no wages at all. Since the early 1990s, Russian workers have been subject to the occasional delay and outright nonpayment of their earned salaries. By the mid- and late-1990s, delays of several months or even years were common. In the fall of 1998, 70 percent of Russian workers reported that they did not get paid regularly, and in about 60 percent of cases, the delays were usually longer than a month. When payment came, it was often after a period of high inflation and thus reflected a lesser value than the original earned wage.[14] Frequent delays and nonpayment of wages have been a principal cause of poverty in Russia, reducing household income, prohibiting spending on more than just basic necessities, exhausting savings, forcing sale of personal property, and increasing reliance on subsistence farming and intra-family transfers of money and food.[15] In several surveys in the 1990s, Russians named wage arrears as the single most serious problem facing their country. The sources of wage arrears are subject to debate. At its peak in 1998, approximately $10 billion worth of unpaid wages could not be accounted for, and there was little agreement on what happened to the money or why workers were not being paid.

Moreover, the demographic breakdown of those most affected by the withholding of their wages is difficult to assess because of the inadequacy of official statistics. The State Statistical Committee, *Goskomstat*, has selectively recorded arrears in certain sectors of the economy and not in others and has been inconsistent in its choices over time. Survey data reveal, however, that patterns in arrears resemble patterns in wage levels. In those regions and industries where wages are already low, workers are most likely to experience wage delays and nonpayment, and, in general, women have been subject to more wage withholding than men.[16]

The problem of wage arrears has not been alleviated by secondary employment any more than the problem of low wages. In surveys conducted in 1994–1997, only 3 percent of working Russians reported that they had a second job.[17] In a 1998 survey, only 10 percent of working Russians reported that they had a second job or second source of income, and this proportion was the same for the unpaid as well as the paid.[18]

Nor have barter or wage substitutes made the lack of wages any more bearable. Although the occasional payment to workers in toilet paper, women's undergarments, coffins, or other outlandish and often insulting goods has gotten much media attention, the incidence of these types of wage substitutes has been much exaggerated. In the first half of 1998, more than a third (35 percent) of unpaid workers had not received goods in lieu of wages even a single time, and just under a third (31 percent) reported receiving goods regularly. Importantly, the magnitude of these payments had only a small mitigating effect on the impoverishing impact of wage arrears.[19]

Inadequate Benefits

The above problems facing labor might not sting so fiercely if Russians could rely on social welfare provisions both at the workplace and in times of unemployment. Unfortunately, however, since the collapse of the Soviet Union, workplace welfare has declined, and the safety net has filled with holes.

Previously, Russian workers could count on their places of employment to provide them with a wide range of basic necessities and luxuries, including housing, medical care, child care, and vacations, and access to hard-to-get consumer goods.[20] The number of firms providing these services has declined steadily, especially for workers who already receive low or no wages.[21] By the fall of 1998, half of workers (51 percent) still reported receiving medical treatment from their workplaces, but many fewer reported receiving other benefits like child care (24 percent), vacations (21 percent), access to a subsidized company cafeteria (18 percent), housing or housing subsidies (11 percent), food (8 percent), or access to a subsidized company store (7 percent). Only a fifth of workers (19 percent) reported receiving two or more of these benefits whether they were unpaid or paid. Moreover, when these benefits came in the form of allowances, payment of the benefits was delayed as often as payment of wages.[22]

As for the unemployed, benefits are either inadequate or unavailable altogether. In a survey conducted in the summer of 1995, two-thirds of respondents said their benefits did not allow them to meet the cost of food.[23] Simon Clarke describes the situation in its starkest terms. "Since the usual level of benefit is considerably below the physiological subsistence minimum, those without additional income and/or the support of family and friends will, by definition, die."[24] Indeed, even having additional sources of income does not fully release the unemployed from the grips of poverty. Over three-quarters of unemployed respondents in the 1995 survey had a per capita monthly income below the subsistence minimum.[25]

Poverty

The end result of these many problems facing workers in contemporary Russia is poverty. Like measures of unemployment, wages, and wage arrears, precise measures of poverty in Russia are subject to debate. Data from *Russian Economic Trends* suggests that prior to the 1998 collapse of the ruble, the share of the population living below the subsistence level was anywhere between 20 and 30 percent, while after the ruble's collapse, the share was in the middle 30th percentile and by June of 2000 had settled back to about 28 percent. Estimates by *Goskomstat* and by survey researchers indicate higher poverty rates in 1996 of 35 and 36 percent, respectively. In general, a reasonable claim is that, since the demise of the Soviet Union, between one-quarter and one-half of the Russian population have been in poverty and that around one-fifth have been in persistent poverty. About one-tenth could be described as living in acute poverty, meaning that per capita income has been below half the subsistence minimum.[26]

Other indicators of poverty corroborate the data on per capita income. For example, a 1996 survey found that two-thirds of households grew some of their own food, and this food was used almost exclusively for personal consumption.[27] (Fewer than 3 percent sold any of the produce.) A 1998 survey found that three-quarters of Russians grew some of their own food, with sizable percentages claiming to grow half (22 percent), more than half (24 percent), or all (9 percent) of the food they ate every day.[28] In 1996, Russians grew 46 percent of all the country's agricultural output (in ruble value) on their household plots, including 90 percent of all potatoes and 75 percent of all vegetables.[29] In short, Russia may be the first nation to deindustrialize, or as some have termed it, demodernize.

Social assistance has not been an effective means of alleviating this widespread poverty. It is estimated that 87 percent of the poor in Russia receive no social assistance at all, and the most poor are the least likely to receive assistance. Social assistance in the country is "pro-rich" in that only 8 percent of social assistance reaches the lowest decile.[30] Poor Russians living in the poorest regions are especially aggrieved, since their regions have the least money to disburse.[31]

Trends in Labor Problems Under Yeltsin and Putin

How did Russian workers come to be in this predicament? In a general sense, the causes of labor problems are easy to identify. They are the same factors that have caused other economic problems in Russia. However, in the more specific case of Russia's most egregious labor problem, wage arrears, the causes are difficult to identify because the problem is multifaceted and complicated.

Problems like unemployment and falling income are the direct result of Russia's decline in production, decline in GDP, capital flight, and limited investment. Entire industries in Russia have proven uncompetitive in market conditions, but few new industries have emerged to replace the old ones and thus create new jobs. In addition, the freeing of prices has launched periods of inflation and even hyperinflation which have sharply eroded the real value of wages, pensions, and state benefits.[32]

Politicians and economists debate the sources of Russia's general economic decline, with some blaming the structural adjustment policies or "shock therapy" implemented during the first years following Soviet collapse and others blaming the poor implementation of these policies. The former claim that shock therapy only encouraged the corruption of well-placed bureaucrats and enterprise managers who defended their own interests at the expense of ordinary workers by, for example, seizing assets. The latter claim that opposition from the immediate victims of reform, like workers, forced a premature retreat from shock therapy, preventing the policies from successfully transforming the economy and raising economic growth and the standard of living.[33]

Regardless of the specific causes of unemployment, low wages, wage arrears, inadequate benefits, and poverty, it is clear that these problems emerged and mounted in the 1990s during the tenure of President Boris Yeltsin. For example, according to data from *Goskomstat*, total employment fell by 8.2 million people from 1990 to 1995, and the percentage of the working age population with full-time employment dropped from 85.5 percent in 1992 to 79 percent in 1995.[34] Survey data reveal differences in the absolute numbers, but they too suggest that unemployment increased markedly during this period.

Most of the problems of Russian labor have improved since Vladimir Putin stepped into the presidency in 2000. However, it remains to be seen whether these improvements are long-lasting thanks to structural changes in the Russian economy or simply a temporary reprieve granted by the increasing price of oil on the world market. Once Putin was in power, the year 2000 saw a 7.5 percent growth in GDP, a nearly 10 percent growth in industrial output, and a nearly 20 percent growth in fixed capital investment.[35] For workers, this meant a corresponding 23 percent increase in average monthly wage, a payoff of arrears to workers on the federal government payroll and to pensioners, and a reduction of arrears to regional government employees and enterprise workers.

Still, many Russians and observers of the Russian scene are skeptical that these initial successes of the Putin regime forecast long-term economic recovery. Structural economic problems remain in the form of aging plants and equipment. Capital flight has continued and even increased, and "Russia continues to depend on world raw-materials prices, which it doesn't control, rather than on the competitiveness of its products."[36] As a result, one-third of the population still lives below the poverty line, average Russian income is still lower than it was before the 1998 ruble collapse, and average living standards have not significantly improved.

Political Implications: The Lack of Protest

Given problems like unemployment and unpaid wages, workers—in Russia or anywhere else—are typically expected to respond politically. According to conventional wisdom, the frustration that accompanies sudden economic decline or sustained poor performance should encourage a rise in demonstrations, strikes, riots, and other forms of civil violence.[37] For example, in January 2001, anger over government corruption and a failing economy led tens of thousands of Filipinos to take to the streets and oust their popularly elected president. In December of that same year, after 42 months of recession in Argentina, nearly 20 percent unemployment, the severe devaluation of savings, the freezing of savings accounts, and the largest public debt default by any country in history, Argentinians also took to the streets. They engaged in riots and strikes, forced the resignation of two successive presidents, and threatened the position of the third and any other would-be successor.

Labor, more than other social groups, is considered especially well-positioned for this type of disruptive behavior. Workers are usually networked through formal unions and the informal ties of daily interaction. They have a clear collective identity and common interests, and, compared to students, neighborhood associations, religious movements, and other groups, they are uniquely situated to wreak economic havoc through work stoppages.[38] For example, during Communist times, it was the labor movement in Poland that threatened to destabilize the poorly performing state socialist regime.

During times of democracy-building, the general fear is that economic frustration might cause such severe public disruption as to provide an opening for extreme political movements on the left and right, provoke military intervention, destabilize the state, and ultimately threaten the chances for successful democratic consolidation. Thus, observers of the Philippines worry that the extra-constitutional path to ousting the democratically elected president has opened the door to further antidemocratic behavior by the new government, and observers of Argentina express concern that military generals might intervene to keep order in the country.[39] The fear is not only that democracy will be derailed but that civil unrest and political instability might discourage investment and thus threaten to derail economic growth and economic reform as well.

Indeed, the 1990s were filled with precisely these types of predictions about the Russian state. If Russia continued to perform poorly economically, it would soon erupt in a "social explosion" or waves of unrest that could threaten the processes of political and economic reform.[40] The initial policies of shock therapy were even predicated on this assumption: Introduce reforms quickly and severely so that the worst pain will be over before frustration can mount and manifest in public opposition.[41] As time went on, the wavering commitment of the Yeltsin regime to radical reform was partly a response to the fear of a public backlash against the pain that reform inflicted. The public backlash was feared to have consequences not only for Russia but for neighboring post-communist regimes as well.[42] When the International Monetary Fund recommended the establishment of social safety nets in Russia, the explicit goal was to counter public opposition to the transition process.[43]

However, despite these many predictions and fears, the public backlash never really came. True, the late 1990s saw periodic outbursts of public indignation, almost exclusively in response to the insidious problem of wage arrears. Teachers, doctors, and coal miners were especially inclined to protest over their long-withheld wages, and regions like Vladivostok and Kemerovo were especially volatile. The forms of protest ranged from standard demonstrations and strikes to more unique disruptive activities like hunger strikes, sit-downs on railroad tracks, and hostage-taking of managers or valuable equipment.

Still, in terms of the quantity of such events and the number of participants, expectations far exceeded reality. Even at the peak of the wage-arrears
crisis, strikers and protesters in Russia represented only 1 or 2 percent of the
population and less than 5 percent of workers actually owed wages. Strikes
and other forms of protest activities were usually not coordinated across regions or sectors of the economy, and they were usually not sustained for very
long. The few attempts at organizing nationwide protest activities, such as in
March of 1997 and April of 1998, usually promised more than they delivered, both in terms of the number and energy of the participants, and they
were forgotten shortly afterwards.

Several observers of Russian politics have attempted to explain this remarkable passivity of Russian workers. Some argue that the dependence of
workers on their places of employment for nonwage benefits like housing
and medical care and for the hope of eventual payment of back wages has
made workers afraid to risk the consequences of protest. Others argue that
participation in Russia's shadow economy has offset the losses from wage
arrears and therefore made protest less necessary. Still others cite the lack of
efficacy among Russian workers and the widespread belief that workers hold
no strategic leverage over their firms to force concessions. Perhaps the most
common argument is that Russian workers lack effective organization. Their
trade unions are incompetent or politically compromised, and other so-called
opposition groups like the Communist Party have not developed a program
to represent workers' interests or rallied workers to action. These and other
arguments may all have some validity, although the evidence seems to support organizational arguments better than arguments about dependency on
the workplace.

Equally or more important is difficulty in attributing blame for complicated economic problems like wage arrears. It might be argued that Russians
have not protested largely because they cannot specify which of the many
potential culprits for their plight is most accountable and therefore most appropriate as a target of action. For example, survey results show that, even
when offered choices of potential culprits, about one-third of the respondents (34 percent) answered "don't know," or they blamed an abstract entity
like the central authorities in general but no one individual or institution in
particular. Just over one-half (54 percent) chose a culprit like President Yeltsin,
the State Duma, or a misguided reform process but were not consistent in
their opinions or in the narrowness of their attributions. For instance, they
might identify the central authorities as most guilty in one question and identify the local authorities in a subsequent question, or they might call multiple
parties "very guilty." Only about one-tenth (11 percent) of Russians were
specific in their attributions of blame.[44] These results are important because

surveys show that the lack of specificity in blame attribution decreases the probability of protest by three to nine times.

Social Implications

In the face of public passivity, most Russian politicians and observers of Russian politics have breathed a sigh of relief. Strikes and demonstrations, after all, could only serve to derail reform and possibly threaten the stability of the Russian state. Worker quiescence is thus generally welcome.

Unfortunately, the complacency or even satisfaction with worker quiescence is ill-advised. Russian workers may not destabilize their political system in the traditional sense of overthrowing the government, but their reaction to their plight takes other forms that threaten the state in less expected ways. Specifically, the political passivity of the Russian public is accompanied by extraordinarily low morale, high stress, and massive-scale depression, and these moods manifest in socially maladaptive behavior like alcoholism, suicide, and crime. The moods and behaviors, combined with the malnutrition that has accompanied poverty, have greatly increased the incidence of disease and mortality in the country. Russia thus faces a public health crisis that is born from its economic crisis and in turn reinforces the economic crisis. The country is trying to reform its economy on a foundation of an increasingly ill and dying workforce, and it is failing to meet the medical costs of caring for that workforce. This state of affairs is hardly cause for complacency, let alone satisfaction.

Stress and Depression

Russian workers have been unable to alleviate their poverty by finding and keeping jobs, raising their wages, ensuring the timely payment of their wages, and ensuring the provision of necessary benefits. They have also been unable to specify a source of blame for all their troubles and thus unable to redress their grievances in the political arena.

In general, Russians lack a sense of efficacy. They "sense that it is not within their power to influence the future direction of their country," and they "feel that politics is beyond their control."[45] For example, the vast majority of working Russians who were deprived of their wages in 1998 (89 percent) believed they could not do anything to help ensure they got paid. The problem, they said, was completely out of their control.[46]

The field of psychology provides some useful terms to characterize the workers' state of mind. These include powerlessness, helplessness, meaninglessness, and loss of control. In general, people want to be self-determining and believe that they are.[47] At times, however, they find themselves in situations where they are not self-determining. They seem to have lost control over their destinies. They feel powerless and helpless, and their lives seem without purpose. It is at these times when high stress and depression, or "psychological distress," are most likely.[48]

Job and income insecurity are the quintessential examples of loss of control. Studies have consistently shown that unemployment is a principal cause of depression.[49] Both the loss of a job and the prospect of becoming jobless increase the frequency of stress and poor mental health, and not just for the unemployed or potentially unemployed. Spouses and children are also affected.[50] The more motivated the unemployed are to find a job, the more deeply they experience the loss of control when they do not succeed, and the more likely they are to suffer depression.[51]

In Russia, the powerlessness caused by job and income insecurity has accordingly led to depression and anxiety on a massive scale. Russians have been described as a demoralized people.[52] In survey after survey conducted in the late 1990s, a solid 40 percent of Russians have described their mood as "tense" and "irritated" while another 10 to 12 percent went so far as to describe themselves as "afraid" and "depressed." Only 4–5 percent described their mood as "excellent" and another 35–40 percent called it "normal" or "balanced."[53] Relations within Russian families have been strained, and pessimism has been widespread. Russians have even come to joke that a pessimist in their country is simply a well-informed optimist.

Alcoholism, Suicide, and Crime

Stress and depression are not just unfortunate mental states; they have dangerous behavioral manifestations. In the United States, unemployment-induced depression has been linked to alcoholism and suicide, with the unemployed being twice as likely to drink heavily and eight times as likely to attempt suicide.[54] Frustration induced by economic deprivation has been linked to murder and other forms of social aggression.[55]

In Russia, the statistics for alcohol abuse, always somewhat worrisome, have reached alarming proportions. Russians drink 4 billion bottles of vodka a year, or nearly forty bottles per adult, and an estimated 20 million Russians, or roughly one-seventh of the population, are alcoholics.[56] A three-year study in Moscow and Udmurtia revealed that two-thirds of Russian men between the ages of twenty and fifty-five who died were drunk at the time of death, even when the immediate cause of death was noted as heart disease, auto accident, suicide, or some other accident or illness.[57]

Drug use born of poverty is also becoming alarming. In the Tuva region alone, where the rate of unemployment is one of the highest in Russia, the number of drug addicts has increased 40 times during the last 10 years, and the number of crimes involving the sale of drugs has risen 10 times. Tuva has 226.1 drug addicts per 100,000 members of the population, compared to the national average of 59.7.[58] Two-thirds (67 percent) of adult Russian males smoke—one of the highest rates in the world.[59]

Russia has also received the dubious honor of becoming the world leader in suicides.[60] In 1990, there were 26.4 suicides per 100,000 people, and by 1994, that number had jumped to 42.1 suicides per 100,000, declining only slightly to 39.8 by 2000.[61] Murders have climbed from 14.3 per 100,000 people in 1990 to 30.6 in 1995. Both suicide and murder have been more prevalent among Russian men than women and seem disproportionately concentrated among the jobless.[62]

In general, the unemployed are more likely to commit crimes. Of those convicted of crimes in Russia, the proportion without employment rose from 20 percent in 1990 to 44 percent in 1995, and this is at a time when total crime and the number of people convicted of crime more than doubled in absolute terms.[63]

Economic Implications

On strictly humanitarian grounds, the situation in Russia is highly troublesome, but it is troublesome from a utilitarian standpoint as well. A stressed, depressed, diseased, and alcohol-abusing population can run up medical costs and costs in criminal justice that put a significant strain on government budgets. Equally important, a stressed, depressed, diseased, and alcohol-abusing workforce is unlikely to make a productive contribution to a struggling economy.

In the West, the medical costs of treating anxiety and depression are enormous. The United States spends $44 billion annually, with an additional $80 billion or so on costs related to alcoholism.[64] In Russia, total costs are unclear, but the Ministry of Health reports that the number of people treated for psychological disorders has risen every year for more than a decade and more than one-third since 1989 so that, by 1999, 3.5 million of Russia's 145 million people were treated for such disorders.

The Health Ministry also estimates that more than one-third of Russians, or 52 million people, have "psychological disorders of various degrees," but given the capacity of Russia's health care system—500 or so mental health institutions able to accommodate only about 200,000 people—many troubled Russians are falling through the cracks and simply not being treated. In general, Russia's health care expenditures as a percentage of GDP are among the lowest in Europe, and Russians with other medical problems are also not being treated.[65] One result is that work hours are lost to psychological and other medical problems, and the nation's productivity suffers.[66]

The main cost to Russia, then, is in allowing these health problems to fester and affect the progress of the country. As Kapstein asserts, "Economic growth simply cannot be sustained in an environment of catastrophic public health."[67] *The New York Times* concurs that "The resulting erosion of the able-bodied work force threatens Russia's hopes for economic revival, and perhaps for political stability as well."[68] Even Prime Minister Mikhail Kasianov has warned that "the problem of the decline of the able-bodied population in the Russian Federation is not simply a social problem. It is a problem . . . of either a successful or an unfavorable development of our state as a whole."[69]

Conclusion: Curing What Ails the Labor Force

Russian workers present a policy challenge for contemporary Russia, but it is not the challenge that the conventional wisdom suggests. Russian workers are unlikely to launch waves of strikes and shut down production in the country or wreak havoc by rioting in the streets, marching on Moscow, or marching on regional governments. What workers are more likely to do is die, and this is no better a circumstance on which to build a thriving democracy and market economy.

Political solutions are challenging because they involve difficult trade-offs. Economic growth depends on a healthy, energetic workforce, which means governments should take care of the unemployed, the ill, and the poor. However, growth also depends on restrained spending and responsible budgeting, which means governments cannot be overly generous in paying out social benefits.

Few specialists on Russia therefore believe that the answer to Russia's labor problems lies in extensive redistribution. Poverty can be cured in a lasting way only through growth. If anything, highly redistributive policies might harm growth by crowding out productivity-enhancing and employment-generating policies.[70] Besides, highly redistributive policies are not very realistic. A 1994 survey revealed that total social assistance in Russia amounted to only 4 percent of the poverty gap and that closing the poverty gap entirely would require Russia to transfer to the poor almost 11 percent of total population expenditures.[71]

Instead, suggestions for curing what ails the labor force in Russia tend to focus on four areas: (1) the promotion of investment and industrial restructuring; (2) legal reforms that protect worker interests; (3) the more effective targeting of social benefits; and (4) the establishment of public works programs.

The first suggestion to improve the situation of labor is the most obvious and the most difficult. Workers need new opportunities, and these can come only from an industrial development strategy that makes investment in the country attractive and thereby promotes economic growth. Job retraining should be a part of this strategy, but it is pointless on its own. Workplace welfare, on the other hand, should not be a part of the strategy. The Soviet tradition of providing housing, food, and other benefits at the workplace only serves to distract firms from concentrating on their primary mission of turning a profit by offering competitive goods and services. The creation of effective housing markets and food distribution systems is a necessary but distinct process from industrial restructuring.[72]

To the extent that workers suffer from arbitrariness from managers, their situation could be improved by legislation that protects their interests and by the ultimate enforcement of such legislation. Workers should be able to rely on the legal system to defend their rights to regular wage payment, and they should be able to bargain collectively to establish their wages and working conditions. They should be guaranteed a minimum wage that is set at least at the subsistence minimum.[73] Some labor specialists go so far as to suggest that workers should be guaranteed retraining and that Russia should restore the former legislation requiring employers to train and place a redundant employee.[74]

While redistribution is not a cure-all for labor problems, neither is it wholly irrelevant. The problem is that redistribution, as it is currently practiced in Russia, is ineffective. In order to alleviate poverty, social expenditures should be targeted to those who are in the greatest need, and expenditures should reflect a realistic assessment of the minimum cost of living. Neither condition currently holds in Russia.

First, as previously mentioned, social assistance in Russia is "pro-rich." Following the social welfare practices of the Soviet state, Russia has been disbursing benefits by category rather than by need. Men and women receive pensions when they reach retirement age, children and full-time students receive child allowances, and the disabled receive benefits, even though many or most in these categories have been demonstrated to be the non-poor.[75] In general, Russia's poor and Russia's non-poor receive social assistance at almost identical rates, and about two-thirds of disbursed social assistance could be considered "leakage," that is, money going to recipients who should not qualify for assistance.[76]

Second, the amount of money that the poor actually do receive per capita is too low to alleviate poverty effectively. Unemployment benefits, for example, are so low that many of the unemployed do not even bother to collect them.[77] Benefits need to be calibrated to the subsistence minimum.

One solution to the problem of ineffective redistribution is in developing instruments to identify the truly poor and redistributing income and benefits accordingly. Grootaert and Braithwaite suggest that the official income test has a high error of exclusion (poor people do not receive benefits) but that a proxy means test using multiple indicators can more accurately distinguish the poor from the non-poor. These indicators can include wage income, car ownership, color television ownership, transfer income, household business, primary education, university education, refrigerator ownership, and land ownership.[78]

Other suggestions include increasing the retirement age. Women currently retire at age fifty-five in Russia. Raising the retirement age to sixty, as it is for men, would allow the country to save 1.5 percent of its GDP. Raising the retirement age further for both men and women would increase savings even further.[79] Of course, raising the retirement age may be politically unviable in Russia, given that well-known decreases in life expectancy heighten public opposition.[80] Russia instead seems to be headed in the reverse direction. It actually increased pensions by 23 percent in 2001.[81]

The situation in contemporary Russia has often been compared to the U.S. Great Depression, and, accordingly, another suggestion to combat the plight of Russian workers is to create programs of public works. Russia could match its great reserves of human capital with the great needs of the country for "maintenance, repair and construction of housing, communal facilities and roads, [and] environmental restitution and improvement."[82] In doing so, it would benefit not only the struggling economy but also the public psyche. It would be handing workers back some control and sense of purpose. Of course, the difficulty in implementing a public works program in an atmosphere of wage arrears and cash deficits is that there is no guarantee that individuals hired for these programs would get paid for their labor.

Still, this particular policy recommendation, more than the others, gets to the heart of Russia's fundamental problem, the lack of empowerment among its people. The Russian labor force faces a material crisis, to be sure, but it also faces a social psychological crisis, and policy needs to address both situations. Russian policy makers need to take the mental and physical health of the Russian public seriously and think of nontraditional approaches to alleviate pervasive feelings of helplessness and loss of purpose. These approaches could come in the economic sphere in the form of public works programs with some greater assurance of wage payment, but perhaps they could also come in the political or civic sphere in the form of civil society development and other forms of community involvement that encourage people to take back control of their destinies.

The probability of Russia's taking the above policy recommendations is mixed. The Putin administration has been devoting considerable energy to courting investment, but it has no program specifically devoted to other worker interests. A new Labor Code passed in December 2001 set the monthly wage to no less than the subsistence minimum, established penalties for managers who delay paying workers' salaries, and otherwise updated Soviet-era labor laws for the first time since Soviet collapse. Unfortunately, however, other legislation like pension reform may actually be counterproductive in continuing to disadvantage the most needy, and labor legislation in general is not a priority. Despite the continued existence of a stagnant pool of the unemployable, public works programs are not even on the agenda, and no one is talking about alleviating the helplessness and depression that plagues the population. All eggs for Russian workers seem to be in the economic growth basket. A little stress seems more than justified.

Notes

I would like to thank John Ambler, Gina Branton, David Brown, Valerie Bunce, Ashley Leeds, Cliff Morgan, Randy Stevenson, Ric Stoll, and Rick Wilson for their helpful comments.

1. Simon Clarke, "Structural Adjustment Without Mass Unemployment," in Simon Clarke, ed., *Structural Adjustment Without Mass Unemployment? Lessons from Russia* (Cheltenham: Edward Elgar, 1998).

2. Ibid., pp. 66, 74.

3. Christian Grootaert and Jeanine Braithwaite, "The Determinants of Poverty in Eastern Europe and the Former Soviet Union," in Jeanine Braithwaite, Christiaan Grootaert, and Branko Milanovic, *Poverty and Social Assistance in Transition Countries* (New York: St. Martin's Press, 2000), p. 49.

4. Galina Monousova, "How Vulnerable Is Women's Employment in Russia?" in Clarke, ed., *Structural Adjustment Without Mass Unemployment?*

5. Clarke, "Structural Adjustment Without Mass Unemployment," in Clarke, ed., *Structural Adjustment Without Mass Unemployment?*, pp.76, 83.

6. Ibid., pp. 46–47.

7. Guy Standing, "Reviving Dead Souls," in Clarke, ed., *Structural Adjustment Without Mass Unemployment?*, p. 184.

8. Clarke, "Structural Adjustment Without Mass Unemployment," in Clarke, ed., *Structural Adjustment Without Mass Unemployment?*, pp. 50–51.

9. Ibid., pp. 53, 77–79.

10. Anders Aslund, "Social Problems and Policy in Postcommunist Russia," in Ethan B. Kapstein and Michael Mandelbaum, eds., *Sustaining the Transition: The Social Safety Net in Postcommunist Europe* (New York: Council on Foreign Relations, 1997), pp. 130–31.

11. Clarke, "Structural Adjustment Without Mass Unemployment," in Clarke, ed., *Structural Adjustment Without Mass Unemployment?*, pp. 78–79.

12. Monousova, "How Vulnerable Is Women's Employment in Russia?" in Clarke, ed., *Structural Adjustment Without Mass Unemployment?*, p. 214.

13. Clarke, "Structural Adjustment Without Mass Unemployment," in Clarke, ed., *Structural Adjustment Without Mass Unemployment?*, pp. 51, 55.

14. See Debra Javeline, *Protest and the Politics of Blame: The Russian Response to Unpaid Wages* (Ann Arbor: University of Michigan Press, forthcoming), introduction.

15. Padma Desai and Todd Idson, *Work Without Wages: Russia's Nonpayments Crisis* (Cambridge: MIT Press, 2000), pp. 211–14.

16. Desai and Idson, *Work Without Wages*, pp. 47–68.

17. Hartmut Lehmann, Jonathon Wadsworth, and Alessandro Acquisti, "Crime and Punishment: Job Insecurity and Wage Arrears in the Russian Federation," *Journal of Comparative Economics*, vol. 27 (1999), pp. 595–617.

18. Javeline, *Protest and the Politics of Blame: The Russian Response to Unpaid Wages*, chapter 5.

19. Desai and Idson, *Work Without Wages*, chapter 11; and Javeline, *Protest and the Politics of Blame*, chapter 5.

20. Stephen Crowley, *Hot Coal, Cold Steel: Russian and Ukrainian Workers from the End of the Soviet Union to the Post-Communist Transformations* (Ann Arbor: University of Michigan Press, 1997), pp. 59–61.

21. Standing, "Reviving Dead Souls," in Clarke, ed., *Structural Adjustment Without Mass Unemployment?*, p. 182.

22. Javeline, *Protest and the Politics of Blame*, chapter 5.

23. Kathleen Young, "Aspects of Official Unemployment in Moscow and St. Petersburg: The Views of the Registered Unemployed," in Clarke, ed., *Structural Adjustment Without Mass Unemployment?*, p. 298.

24. Clarke, "Structural Adjustment Without Mass Unemployment," in Clarke, *Structural Adjustment Without Mass Unemployment?*, p. 65.

25. Young, "Aspects of Official Unemployment in Moscow and St. Petersburg: The Views of the Registered Unemployed," in Clarke, ed., *Structural Adjustment Without Mass Unemployment?*, p. 298.

26. Simon Clarke, "Poverty in Russia" (Department for International Development, December, 1997). Available at www.warwick.ac.uk/fac/soc/complabstuds.russia/Poverty_Russia.doc.

27. Clarke, "Structural Adjustment Without Mass Unemployment," in Clarke, ed., *Structural Adjustment Without Mass Unemployment?*, p. 80.

28. Javeline, *Protest and the Politics of Blame*, chapter 5.

29. Clarke, "Structural Adjustment Without Mass Unemployment," in Clarke, ed., *Structural Adjustment Without Mass Unemployment?*, p. 80.

30. Branko Milanovic, "The Role of Social Assistance in Addressing Poverty," in Jeanine Braithwaite, Christiaan Grootaert, and Branko Milanovic, *Poverty and Social Assistance in Transition Countries* (New York: St. Martin's Press, 2000), pp. 130, 133.

31. Jeanine Braithwaite, Christiaan Grootaert, and Branko Milanovic, "Policy Recommendations and General Conclusions," in Braithwaite, Grootaert, and Milanovic, ibid., p. 168.

32. Michael Mandelbaum, "Introduction," in Kapstein and Mandelbaum, eds., *Sustaining the Transition*, p. 2; and Grootaert and Braithwaite, "The Determinants of Poverty in Eastern Europe and the Former Soviet Union," in Braithwaite, Grootaert, and Milanovic, *Poverty and Social Assistance in Transition Countries*, p. 47.

33. Clarke, "Structural Adjustment Without Mass Unemployment," in Clarke, ed., *Structural Adjustment Without Mass Unemployment?*, pp. 11–13.

34. Standing, "Reviving Dead Souls," in ibid., p. 150.

35. Otto Latsis, "Good, but Not Good Enough," *The Russia Journal*, January 13–29, 2001, reprinted in Johnson's Russia List, no. 5029, January 15, 2001. Available at www.cdi.org/russia/johnson.

36. Ibid.

37. Leslie Elliott Armijo, Thomas J. Biersteker, and Abraham F. Lowenthal, "The Problems of Simultaneous Transitions," *Journal of Democracy*, vol. 5, no. 4 (1994), p. 164; and Stephen Haggard and R. R. Kaufman, *The Political Economy of Democratic Transitions* (Princeton: Princeton University Press, 1995), p. 334.

38. J. Samuel Valenzuela, "Labor Movements in Transitions to Democracy: A Framework for Analysis," *Comparative Politics*, vol. 21, no. 4 (July 1989), p. 447.

39. *The Economist*, January 17, 2002, and ibid., February 28, 2002. Available at www.economist.com.

40. Adam Przeworski, *Democracy and the Market: Political and Economic Reforms in Eastern Europe and Latin America* (New York: Cambridge University Press, 1991), p. 189; Adam Przeworski, et al., *Sustainable Democracy* (New York: Cambridge University Press, 1995), p. 110; Mandelbaum, "Introduction," in Kapstein and Mandelbaum, eds., *Sustaining the Transition*, p. 3; and Katrina vanden Heuvel and Stephen F. Cohen, "The Other Russia," *The Nation*, August 11–18, 1997.

41. Anders Aslund, "The Case for Radical Reform," *Journal of Democracy*, vol. 5, no. 4 (1994), p. 169.

42. Mandelbaum, "Introduction," in Kapstein and Mandelbaum, eds., *Sustaining the Transition*, pp. 3, 6.

43. Ethan B. Kapstein, "Toward a Political Economy of Social Policy," in Kapstein and Mandelbaum, eds., *Sustaining the Transition*, p. 179.

44. Javeline, *Protest and the Politics of Blame*, chapter 5.

45. Ashwin, *Russian Workers: The Anatomy of Patience*, p. vii.

46. Javeline, *Protest and the Politics of Blame*, chapter 5.

47. Robert E. Lane, *The Loss of Happiness in Market Democracies* (New Haven: Yale University Press, 2000), p. 231.

48. Richard Nisbett and Lee Ross, *Human Inference: Strategies and Shortcomings of Social Judgment* (Englewood Cliffs: Prentice-Hall, 1980); and John Mirowski and Catherine E. Ross, *Social Causes of Psychological Distress* (New York: Aldine de Gruyter, 1989).

49. Michael Argyle, *The Social Psychology of Everyday Life* (New York: Routledge, 1992).

50. Lane, *The Loss of Happiness in Market Democracies*, p. 166.

51. Norman T. Feather and P. R. Davenport, "Unemployment and Depressive Affect: A Motivational and Attributional Analysis," *Journal of Personality and Social Psychology*, vol. 41 (1981), pp. 422–36.

52. Clarke, "Structural Adjustment Without Mass Unemployment," in Clarke, ed., *Structural Adjustment Without Mass Unemployment?*, pp. 51, 83.

53. Oleg Savelyev, "Subsistence Optimism," *Vremia MN*, February 11, 2000, reprinted in Johnson's Russia List, no. 4108, February 15, 2000. Available at www.cdi.org/russia/johnson.

54. Argyle, *The Social Psychology of Everyday Life*, p. 264.

55. Lewis A. Coser, *Continuities in the Study of Social Conflict* (New York: The Free Press, 1967), pp. 56–62; and Ted Robert Gurr, *Why Men Rebel* (Princeton: Princeton University Press, 1970), pp. 73–74.

56. Murray Feshbach, "Russia's Population Meltdown," *Wilson Quarterly*, vol. 25, no. 1 (Winter 2001), pp. 15–21.

57. Julie A. Corwin, "Study Finds Most Men Die Drunk," *RFE/RL Newsline*, vol. 4, no. 98, Part I, May 22, 2000. Available at www.rferl.org/newsline.

58. Julie A. Corwin, "Drugs Transforming Tuva," *RFE/RL Newsline*, vol. 5, no. 3, Part I, January 5, 2001. Available at www.rferl.org/newsline.

59. "Highlights on Health in the Russian Federation," World Health Organization, November, 1999, p. 19. Available at www.who.dk.document/e72504.pdf.

60. Corwin, "Study Finds Most Men Die Drunk."

61. Paul Goble, "2000 Mortality Figures Worse than 1990, Better than 1995," *RFE/RL Newsline*, vol. 5, no. 10, Part I, January 16, 2001. Available at rferl.org/newsline.

62. Standing, "Reviving Dead Souls," in Clarke, ed., *Structural Adjustment Without Mass Unemployment?*, pp. 154–55.

63. Ibid., p. 156.

64. Lane, *The Loss of Happiness in Market Democracies*, p. 329.

65. "Highlights on Health in the Russian Federation," World Health Organization, November, 1999, p. 19. Available at www.who.dk.document/e72504.pdf.

66. *Chicago Tribune*, February 1, 2001.

67. Kapstein, "Toward a Political Economy of Social Policy," in Kapstein and Mandelbaum, eds., *Sustaining the Transition*, p. 183.

68. *The New York Times*, January 2, 2001.

69. *BBC Monitoring*, February 15, 2001, reprinted in Johnson's Russia List, no. 5097, February 16, 2001. Available at www.cdi.org/russia/johnson.

70. Kapstein, "Toward a Political Economy of Social Policy," in Kapstein and Mandelbaum, eds., *Sustaining the Transition*, pp. 181, 184; Mandelbaum, "Introduction," in ibid., p. 5; and Scott Thomas, "Social Policy in the Economies in Transition: The Role of the West," in ibid., 152–54.

71. Milanovic, "The Role of Social Assistance in Addressing Poverty," in Braithwaite, Grootaert, and Milanovic, *Poverty and Social Assistance in Transition Countries*, pp. 128–29.

72. Mandelbaum, "Introduction," in Kapstein and Mandelbaum, eds., *Sustaining the Transition*, p. 5.

73. Aslund, "Social Problems and Policy in Postcommunist Russia," in Kapstein and Mandelbaum, eds., *Sustaining the Transition*, p. 133; and Simon Clarke, Veronika Kabalina, Irina Kozina, Inna Donova, and Marina Karelina, "The Restructuring of Employment and the Formation of a Labour Market in Russia," in Clarke, ed., *Structural Adjustment Without Mass Unemployment?*, p. 145.

74. Clarke, et al., "The Restructuring of Employment and the Formation of a Labour Market in Russia," in Clarke, ed., *Structural Adjustment Without Mass Unemployment?*, p. 145.

75. Aslund, "Social Problems and Policy in Postcommunist Russia," in Kapstein and Mandelbaum, eds., *Sustaining the Transition*, pp. 142–43; Grootaert and Braithwaite, "The Determinants of Poverty in Eastern Europe and the Former Soviet Union," in Braithwaite, Grootaert, and Milanovic, *Poverty and Social Assistance in Transition Countries*, pp. 85–86.

76. Milanovic, "The Role of Social Assistance in Addressing Poverty," in Braithwaite, Grootaert, and Milanovic, *Poverty and Social Assistance in Transition Countries*, p. 126.

77. Aslund, "Social Problems and Policy in Postcommunist Russia," in Kapstein and Mandelbaum, eds., *Sustaining the Transition*, p. 142.

78. Grootaert and Braithwaite, "The Determinants of Poverty in Eastern Europe and the Former Soviet Union," in Braithwaite, Grootaert, and Milanovic, *Poverty and Social Assistance in Transition Countries*, pp. 85–90.

79. Aslund, "Social Problems and Policy in Postcommunist Russia," in Kapstein and Mandelbaum, eds., *Sustaining the Transition*, p. 142.

80. Ibid., p. 184.

81. Paul Goble, "Russia to Augment Planned Pension Increase," *RFE/RL Newsline*, vol. 5, no. 16, Part I, January 24, 2001. Available at www.rferl.org/newsline.

82. Clarke et al., "The Restructuring of Employment and the Formation of a Labour Market in Russia," in Clarke, ed., *Structural Adjustment Without Mass Unemployment?*, pp. 145–46.

For Further Reading

Stephen Crowley, *Hot Coal, Cold Steel: Russian and Ukrainian Workers from the End of the Soviet Union to the Postcommunist Transformations* (Ann Arbor: University of Michigan Press, 1997).

Ethan B. Kapstein and Michael Mandelbaum, eds., *Sustaining the Transition: The Social Safety Net in Post-Communist Europe* (New York: Council on Foreign Relations, 1997).

Padma Desai and Todd Idson, *Work Without Wages: Russia's Nonpayments Crisis* (Cambridge: MIT Press, 2000).

Debra Javeline, *Protest and the Politics of Blame: The Russian Response to Unpaid Wages* (Ann Arbor: University of Michigan Press, forthcoming).

11

Russia's Demographic Challenges

Timothy Heleniak

At the beginning of the twenty-first century, Russia confronts a number of demographic challenges. Since Russia became independent in 1992, the combination of declining fertility, increasing mortality, an aging population, and reduced levels of immigration has caused the country's population to begin to decline. While Russia's population has fallen by only 3 percent, the rate of decline has begun to accelerate in the past few years and is expected to continue to the plausible limit of demographic projections. This situation has alarmed the country's leadership, who worry about the effects of this decline on Russia's status as a superpower, and Russian nationalists who fear—probably prematurely—the demise of the Russian nation. Qualitatively, the health of the Russian population—already low by international standards—has worsened significantly during the transition period. Russia's highly trained labor force has shown signs of deteriorating, as access to educational places is no longer guaranteed and many of the country's best and brightest have left. The rapid aging of the Russian population, caused in part by the steep fertility decline, has made caring for the country's elderly population an increasing financial burden.

In addition, the breakup of the Soviet Union, the transition away from the centrally planned economy, and the institution of a democratically elected leadership have influenced the three components of demographic change—fertility, mortality, and migration. Taken together, Russia faces one of the most critical demographic challenges in its history. The purpose of this chapter is to analyze the demographic challenge confronting Russia. The section below reviews demographic trends leading up to the fall of the USSR.

Demographic Trends: Births, Deaths, and Fertility

Starting with World War I in 1914 through the end of the second World War in 1945, the Russian population was subjected to a series of events that would leave an indelible mark on its population size, distribution, and age-sex structure. These included World War I, the Bolshevik Revolution and subsequent Civil War, the famine of 1932–33 caused by the collectivization of agriculture, the purges of 1934–36, and World War II.

The total demographic losses over the 1927-to-1946 period are estimated at 56 million. Adding births that didn't take place, the Russian population could have grown by another 80 million persons over this period. Had the various demographic disasters of the Soviet period not taken place, the Soviet Union would have had a population of 440 million at the end of its existence in 1991 rather than the 286 million it did have.

In spite of the various demographic disasters of the first half of the twentieth century, Russia continued its demographic transition from high birth and death rates to lower birth and death rates. At the beginning of the twentieth century, Russian women were, on average, giving birth to more than seven children, due partially to the absence of reliable fertility control.[1] As a result of the rapid industrialization and urbanization of the Soviet period, Russia's fertility levels rapidly converged toward those in the rest of Europe. By the early 1960s, the fertility of Russian women had fallen into a rather stable two-child pattern, with the total period fertility rate falling below 2.5 children per woman for the first time in 1960.

The package of social support provided by the Soviet state did serve as an incentive to have children. This consisted of free primary, secondary, and higher education, free medical care, guaranteed employment, housing for families with children,[2] and ready availability of child-care facilities at most workplaces. As a result, Russian women married and had children early, and both activities were nearly universal.[3] Further, life expectancy rose by 28 years for males from 1927 to 1958 from 33 years to 62 years and for females by 33 years from 38 years to 70 years. By the early 1960s, life expectancy for Russians was approaching that of the United States. The life expectancy gains then stalled for the most part because the burden of disease shifted from infectious diseases that had been brought down considerably to more lifestyle diseases that the Soviet healthcare system was ill-equipped to prevent.

Partially influenced by the package of pro-natalist measures, the number of births in Russia peaked in 1987 at 2.5 million (see Figure 11.1). Since then, the number of births has fallen by 48 percent, to just 1.3 million in 2001. The same period, 1985 to 1988, was the last time that the Russian fertility rate was above replacement level, peaking at 2.194 in 1986–1987. It fell by over one child per woman since to 1.171 in 1999 before recovering slightly in 2000 to 1.214. This places Russia, along with a number of other transition countries and several developed countries in western Europe, in a category of countries with the lowest fertility rates in the world.[4] This steep

Figure 11.1   **Russia: Number of Births and Total Fertility Rate, 1950–2001**

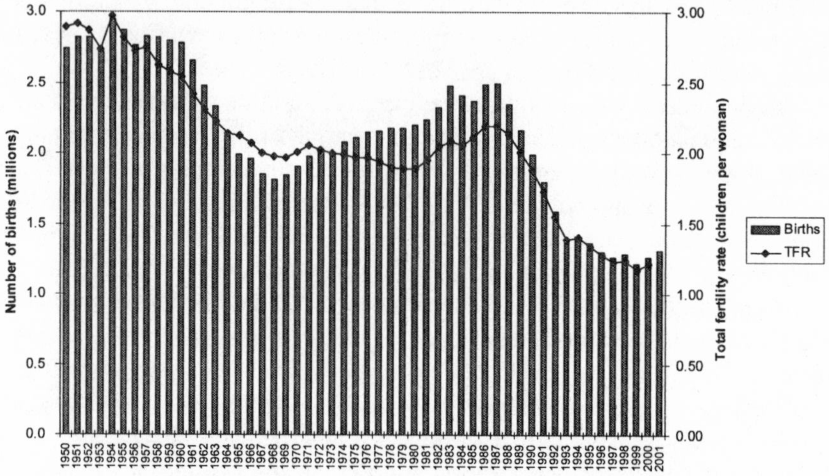

Source: Goskomstat Rossii, *Demograficheskii ezhegodnik* (various years).

drop in the number of births has had a greater numerical impact on the population decline in Russia than the rising death rate, which has attracted so much attention. It is the fertility rate, more so than the mortality rate, that the Russian government has as a demographic lever to influence the country's population size. Many casual observers have looked at the fertility decline in Russia and the overall population decline and attributed it to the social crisis caused by the economic transition and breakup of the Soviet Union. While these events have contributed to Russia's current negative demographic trends, that is not the entire explanation. There are factors in Russia's demographic past that have contributed to the fertility decline as well as more difficult-to-quantify factors such as the transition to a more Western pattern of child bearing as a result of the reduction of state social protection.

The rapid decline in the birth rate over the past decade will cause further kinks in Russia's already peculiar age pyramid, further complicating economic and social planning. The fertility decline has contributed to an already aging Russian population. Overall in Russia, there was an 11 percent increase in the elderly population from 1989 to 2000. The percent of the population above age 60 is expected to increase from 18.5 percent in 2000 to 26.0 percent in 2025, and further to 37.2 percent by 2050. This will place further strains on Russia's pay-as-you-go pension system. Later in this decade, the number of new pensioners will begin to greatly exceed the number of new entrants to the labor force. The rapid aging of the population will force Russia to spend large amounts on pensions and other benefits for the elderly. On the other hand, it does provide something of a window of opportunity, as the number of school-age children has declined by 8 million over the same period, allowing some breathing room for necessary education reforms.

Many observers have looked at the intersection of crude birth and death rates, noted that they intersect in 1992, the year of the breakup of the Soviet Union, and labeled the phenomena "the Russian cross." While the situation that Russia (as well as many other transition states) now finds itself in—with more deaths than births—started in the first year of its independence and the start of its economic transition, the two events are coincidental but not necessarily correlated. The current and future shape of Russia's population profile are embedded in its tortured demographic past. The negative natural increase (or natural decrease) was expected to start early in the twenty-first century regardless of the breakup of the Soviet Union and the market transition because of Russia's age structure: fewer people entering childbearing age than were entering older age with higher mortality rates.[5] The transition accelerated this trend, mainly by causing the birth rate to plunge but also by increasing the death rate. There are a number of the transition countries as well as some of the older West European countries where deaths exceed births because of a combination of aging populations and decades of below-replacement fertility. Russia is distinguished from these countries in two respects. First, of the approximately fourteen countries where deaths exceed births,[6] only Ukraine has a higher *rate* of natural decrease. Second, given Russia's large population size, the amount of natural decrease is enormous. During each of the past three years, the excess of deaths over births has been nearly 1 million.

The fact that there has been a dramatic decline in the number of births in Russia during the transition period is irrefutable. The crucial issues for Russia's demographic future are the causes of this decline and whether this is evidence of a short-term decline in response to the economic depression of the 1990s or the beginning of a long-term shift away from the two-child norm. More illuminating is data on the number of births by birth order. The overall fertility rate in Russia over the past decade has been driven by the second-birth fertility rate.[7] Many Russian women responded to the pro-natalist incentives of the 1980s and had their first or second births "ahead of schedule," which partially contributed to the fall in the fertility rate in the early 1990s.[8] During the 1990s, there has been a decline in the fertility rate of the second and higher birth orders but not in the first. As a result, first births went from constituting 49.5 percent of all births in 1990 to 58.8 in 1998. This would seem to indicate that, at least during the troubled period of 1990s, women were tending to limit family size to one.

Will they return to the two-child norm when the economic situation improves? It's too early to tell, but the slight rise in the fertility rate in 2000 was partially due to women 30 years and older increasing their fertility rates more than the average for all women in their childbearing years. However, in a study of the causes of low fertility among developed countries, it was found that the total period fertility rates (the number of children being born to the current hypothetical cohort of women in their childbearing years) are temporarily depressed due to a rise in the mean age at childbearing and that the distortion can be as much as 0.4 births per woman in the lowest-fertility countries such as Italy and Spain.[9] However, these temporary effects will eventually stop depressing fertility in all countries except one, Russia. Thus, Russia among developed countries cannot expect a boost in fertility in the future from seemingly invisible temporary effects in the timing of childbearing. Additionally, the effects of the smaller cohorts of the past 13–14 years on future fertility should be kept in mind. Starting in 2007, the number of women in their childbearing years will decline significantly, from 12.7 million in that year to 7.2 million in 2022. Thus, Russian women would have to increase their fertility levels significantly to compensate for their smaller cohorts and to stop the Russian population from declining.

Factors Affecting Demographic Trends

Research has shown that there are four variables that explain most of the differences in fertility among populations—proportion of females who are married, contraceptive prevalence, the abortion rate, and the duration of postpartum infecundabilty due to breast-feeding.[10] In Russia, during the transition period of the 1990s, these proximate determinants of fertility all moved in a direction to drastically lower the fertility rate. Surveys have shown that Russian women tend not to breast-feed for very long after childbirth, so this is not a major factor in explaining fertility levels. The availability of contraceptives in Soviet Russia was limited, and those that were available have been described as being of poor quality.[11] Contraceptive prevalence rates (CPR) of 34.8 percent were estimated for 1979 and 31.4 percent for 1988. Following the breakup of the Soviet Union, imported contraceptives became more readily available, often, however, at rather high prices. In some cases, international donors sought to provide modern contraceptives at little or no cost specifically to reduce the reliance on abortions. As a result, the CPR more than doubled to 66.8 percent in 1994, with about half of those using modern methods employing intrauterine devices (IUDs). There are problems with the data on contraceptive use, but, regardless of the precise level, they show a very rapid rise in one of the major fertility inhibiting variables in Russia during the 1990s.

Similar to data on contraceptive use, there are problems with abortion data.[12] Thus, analyzing times series of abortion rates and ratios will include an undeterminable amount of statistical omission. According to official data, the number of abortions in Russia fell by over half from 1989 to 2000.[13] The abortion ratio has been near or above 200 abortions per 100 births for most of the three decades prior to the breakup of the Soviet Union. Since 1989, the officially reported ratio has fallen from just over 200 abortions per birth to 169. In spite of possible measurement problems, the figure is extremely high, keeping Russia among the highest in the world. Thus, even if the use of abortion as a means of fertility regulation has declined, its use remains widespread, further explaining the fertility decline in post-Soviet Russia. Further pushing down fertility as a result of abortion is the impact of high abortion rates on secondary sterility of women.

A third major variable affecting fertility is the proportion of females who are married. During the Soviet period, first sexual contact, first marriage, and first pregnancy tended to coincide.[14] Marriage was nearly universal, with only 3.7 percent of males and 3.4 percent of women 45–49 years of age not having been married, and childlessness was rare. A census is required to truly determine the proportion married, but all indications are that the proportion of women married has declined considerably. Since the transition began, marital rates have declined, and marriage and pregnancy have become independent. Since 1987, the number of marriages has declined by 31 percent, while the number of divorces has increased by 32 percent over the same period. In 1950 only 4 percent of Russian marriages ended in divorce. This figure climbed to around 40 percent in the 1980s and jumped to 76 percent in 2001. Thus, one of the strongest links to first pregnancies has declined considerably during the 1990s. This has been partially compensated for by an increase in births taking place outside of marriage. The share of births outside of marriage increased from 13 percent in 1988 to 28 percent in 2000. However, the increase in non-marital fertility is hardly compensating for the decline in marital fertility. The number of births to married women dropped from 1,868,834 in 1989 to 912,547 in 2000, a decline of over 900,000. The number of non-marital births went from 291,725 to 354,253 over the same period, an increase of only 62,528.[15] Of those births outside marriage, a rather stable share of 41–47 percent was to cohabitating couples.[16] The rest were presumably to single mothers where the father was not involved. This provides some evidence that Russia is undergoing what is referred to a second demographic transition, long under way in the West, that is characterized by an increase in consensual unions, a larger contribution of births outside marriage, an increase in average age at marriage, first birth, and all births, and a later peak of childbearing.

Among other factors contributing to the decline in the birthrate is the explosion in the number of sexually transmitted diseases. The official reported number of newly registered cases of STDs has nearly doubled from 1989 to 2000, from 209,000 cases to 415,000.[17] A widespread prevalence of venereal diseases can cause a high prevalence of sterility, further lowering a population's fertility, and figures have been given stating that 5 million women and 3 million men in Russia are infertile.[18]

Until now, HIV/AIDS has not constituted a very large portion of the overall burden of disease in Russia, with only 699 AIDS deaths.[19] However, the rate of increase in prevalence is rising faster in Russia, and elsewhere in the former Soviet Union, than anywhere else in the world, according to the officially reported numbers. These are widely acknowledged to underreport the true prevalence of the epidemic, with the actual levels being three, five, or ten times higher. Part of the reason for the under-enumeration of the infected population in Russia was the reliance on mass blood-screening of the population rather than targeting high-risk groups. If the actual number is closer to 1 million, which is the consensus figure, this would indicate a prevalence rate roughly double that of the United States. Of the 193,400 officially registered cases of HIV-AIDS in Russia (as of May 1, 2002), 94 percent have occurred since 1999. The epidemiological trajectory of HIV/AIDS in Russia differs substantially from other countries. The prevalence of HIV in Russia was rather low by international standards, with only 1,072 registered cases through the end of 1995, the majority attributed to homosexual or heterosexual contact. A new trend emerged in the mode of transmission in 1996 that has helped to fuel the extremely high incidence rates, when injecting drug users (IDU) became widely involved in the epidemic through sharing of contaminated needles. The share of new cases attributable to IDU went from virtually zero in 1993 to nearly 90 percent in 1999. At the end of 2001, those infected are predominantly young males, 78 percent of the total and 62 percent of male cases being between ages 20 and 30.[20] There are currently one-half million registered intravenous drug users in Russia with estimates of the true number ranging from 2.5 million to 4 million. There is fear and evidence that from this sexually active and high-risk population a more generalized HIV/AIDS epidemic is emerging in Russia that could grow rapidly and cripple economic development. HIV is often a co-infection with TB, the rates of which are also increasing rapidly and which are expected to increase even further with the amnesty of prisoners in Russia. TB incidence rose from 55,469 in 1989 to 130,685 in 2000.[21] With the lack of success of DOTS and other programs that have proven effective elsewhere in the world, the number of TB deaths has increased from 13,784 in 1992 to over 29,000 the past two years.[22]

The various apocalyptic predictions for the possible increase in HIV/AIDS mortality are not reassuring. A recent World Bank study estimated the economic consequences of the epidemic in Russia under several different scenarios. The economic consequences are the declines in the numbers of persons in the labor force through an increase in deaths, declines in productivity among HIV-infected workers, and diversion of funds, either public or private, for treatment. The annual cost of treating an HIV-positive person in Russia has been estimated at $3,000. The cumulative number of HIV-positive persons in 2020 ranges from 5.36 million to 14.53 million persons, making the cost of treatment astronomical. The annual number of HIV/AIDS-attributable deaths in 2020 ranges from 250,000 to 650,000. Unlike countries in sub-Saharan Africa with rather high birth rates where the epidemic is slowing the rate of growth, if these projections for Russia prove true, they would occur amongst a population that is already declining by nearly a million persons annually from natural decrease. Without effective policies or cost-effective treatment programs, GDP would be 4.2 percent lower in 2010 and 10.6 percent lower in 2020 as a result of the epidemic. Thus, left unchecked the HIV/AIDS epidemic could have dire consequences for all aspects of Russian society.

There is considerable agreement among demographers both inside and outside the country that the psychosocial stress brought on by the transition contributed to the steep decline in life expectancy in the 1990s.[23] Compared to the late 1980s, male life expectancy fell by 7.3 years in 1994, while female life expectancy fell by less than half of that, 3.4 years, at which time the gender gap was 13.6 years. Life expectancy for both sexes had recovered about half of the decline to 1998, and the gender gap had narrowed to 11.6 years. In 2001, following the 1998 ruble crisis, life expectancy for males had fallen to 58.6 years and for females to 72.1 years, with the gap increasing 13.5 years, about the level at the depth of the Russian mortality crisis. Thus, it appears that males in Russia are far more susceptible to economic and social dislocations than females. Figure 11.2 shows life expectancy by gender in Russia.

In addition to reduced life expectancies, an increase in mortality during the 1990s affected those in the younger middle ages as is shown in Figure 11.3. The largest increase has been among males ages 20–24, with significant increases among males up to age 44. At ages less than 20 and more than 44, increases in death rates were much less, with death rates at the extremes even showing improvements. For females, the same broad pattern emerges of a peak among 20- to 24–year-olds and smaller increases at ages younger and older than this. For males, three-quarters of the increase in death rates from 1989 to 1994, during the time of the

Figure 11.2   **Russia, Life Expectancy by Sex, 1960–2001**

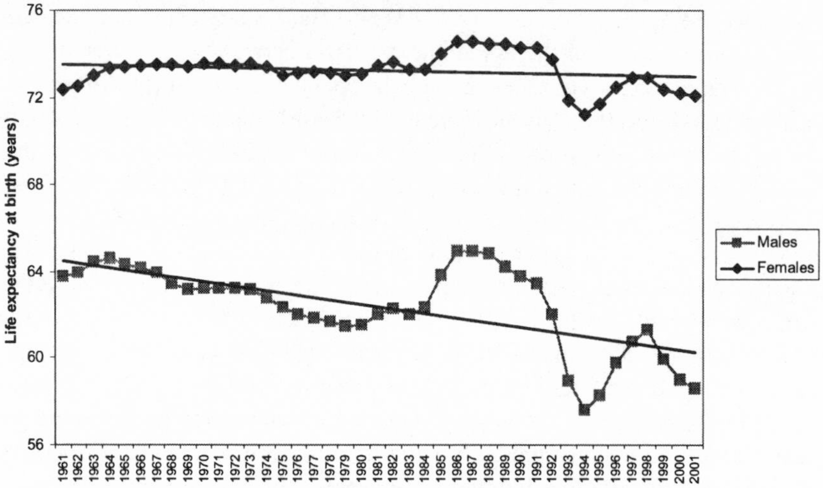

Source: Goskomstat Rossii, *Demograficheskii ezhegodnik* (various years).

steepest declines in life expectancy, can be attributed to increases in two major causes of death. Forty-two percent is attributable to increases in diseases of the circulatory system and 33 percent to trauma and poisonings (see Figure 11.4).[24] For females, these same causes contributed to the increases in the standardized death rates in roughly the same proportions (46 percent of the increase to diseases of the circulatory system and 20 percent to trauma and poisonings), although the percent increase in the death rate for females was much lower than for males. Likewise, during the period of increase in life expectancy from 1994 to 1998, diseases of the circulatory system (46 percent) and deaths from trauma and poisoning (25 percent) other accounted for 80 percent of the decline among males (respectively 59 and 17 percent among females). Cardiovascular diseases explain much of the differences in life expectancy between Russia and Western Europe and thus are the primary cause of long-term fluctuations in mortality in Russia. Injuries and poisonings primarily explain short-term fluctuations, especially among males.[25] The Russian mortality situation is not helped by the fact that 63 percent of males and 13 percent of females are smokers, among the highest smoking prevalence rates in the world.[26] Thirty percent of all deaths in Russia can be attributed to smoking, including 52 percent of all cancers and 29 percent of all cardiovascular diseases. Though there have been some promising anti-smoking measures taken recently in Russia, it will likely take decades before these measures have any appreciable effect, if the example of the

Figure 11.3   **Russia: Changes in Age-Specific Mortality Rates, 1990–2000**

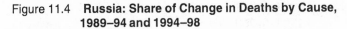

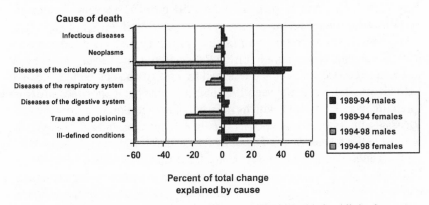

Source: Goskomstat Rossii, *Demograficheskii ezhegodnik* (various years).

Figure 11.4   **Russia: Share of Change in Deaths by Cause, 1989–94 and 1994–98**

Source: From Brainerd and Varavikova based on WHO Health for All database.

anti-smoking campaign in the United States is any indicator. Any Russian anti-smoking efforts also have to combat the recent "westernization" of the Russian tobacco industry.

International Migration Trends

The breakup of the Soviet Union into fifteen independent states left 25.2 million Russians living outside Russia. Some, but not all of international migration during the past decade has been the "return" of diaspora groups, including Russians, to their ethnic homeland. The breakup of the Soviet Union has also brought many long-suppressed ethnic conflicts into the open. With the exception of the ongoing conflict in Chechnya, most of this has not been aimed at Russians but they have been caught in the middle of much of it and have been forced to flee as a result. As many of the non-Russian states have chosen to promote the titular group over Russians, this has acted as a push factor to the Russian groups in these states. The problem for Russia has been that for much of the 1990s, "returning" migration into Russia compensated for declining birthrates and increasing death rates. However, as the decade wore on, the volumes of in-migration from the near abroad lessened, bringing into question the viability of depending on migration to make up for negative natural increases.

For most of the Soviet period, there was net out-migration from Russia to the non-Russian states. This trend was reversed in 1975, and from that year until the breakup of the Soviet Union at the end of 1991, net migration into Russia from the non-Russian FSU averaged about 160,000 annually. Net migration into Russia rose rapidly following the breakup of the Soviet Union, peaking at 809,614 in 1994. In 2001, there was return migration to Russia of only 72,300, less than 10 percent of the 1994 peak. The levels of migration with the other successor states largely drive the overall levels of net migration in Russia, as migration with the far abroad has remained rather steady over the decade. Net emigration from Russia to the far abroad has averaged just about 100,000 annually, far less than flows many had predicted once exit barriers were removed. Between 1989 and 2000, the population increase from migration was 3.6 million, which consisted of net immigration from the non-Russian states of the former Soviet Union of 4.7 million and a net emigration to outside the former Soviet Union of 1.1 million.

The patterns of migration by country for Russia since 1989, with both the FSU states and the far abroad, have largely been driven by the nationality composition of those migration streams (see Figure 11.5). Three countries account for the bulk of persons migrating from Russia to beyond the former Soviet Union—Germany with about 57 percent, Israel with 26 percent, and the United States with 11 percent. Because of this geographic concentration of migration, the flows to the far abroad consisted primarily of three groups—Germans, Jews, and Russians. Since 1989, Russia has had a positive migration balance every year with all of the other FSU states, with the exception for some years of Belarus and Ukraine. Between Russia and the non-Russian states, the three states with the largest Russian diaspora populations account for the largest shares of immigration between 1989 and 2000—Ukraine and Kazakhstan each with a quarter of total immigrants and Uzbekistan with 11 percent. Overall, Central Asia has been the source for about one-half of all migrants to Russia, the three Transcaucasus states 15 percent, the Baltics only 4 percent.

Figure 11.5 **Net Migration by Country to Russia, 1989–2001**

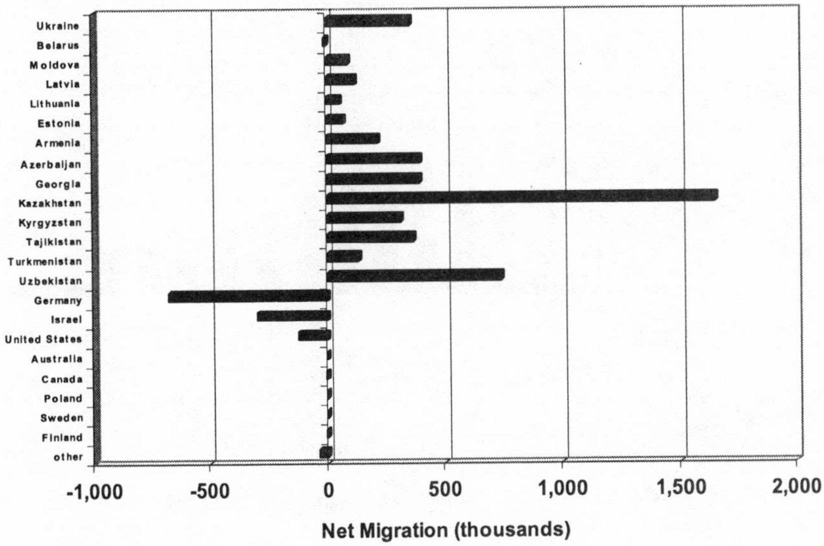

Source: Goskomstat Rossii, *Demograficheskii ezhegodnik* (various years).

A clear regional grouping emerges in terms of the percentages of Russians residing in the non-Russian states of the former Soviet Union who have left each of the newly independent states (see Figure 11.6). From four states, Armenia, Tajikistan, Azerbaijan, and Georgia, one-half or more of the Russian populations have chosen migration as a strategy of adaptation. It was also from these states that significant shares of the titular populations have fled as well because of episodes of ethnic violence that have caused deteriorating economic conditions. Tajiks, Armenians, Georgians, and Azeris all increased their size in Russia by significant numbers as a result of migration. From two states, Uzbekistan and Kyrgyzstan, roughly a quarter of the Russian populations have left. Kazakhstan makes up its own group in terms of the share of Russians who have left (17.8 percent), while being the state from which the largest absolute numbers of Russians have migrated. A fourth group is the three Baltic states and Moldova where between 10 and 13 percent of the Russian diaspora populations have left. A fifth group are the other two Slavic states where only small portions of Russians have returned.

During the chaotic period from 1989 to 1993, while the Soviet Union was in the process of dismantling, some of the titular nationalities of the non-Russian successor states had net emigration from Russia for some years and net immigration for others with no clear pattern emerging. Only the Arme-

Figure 11.6 **Net Migration of Russians into Russia, 1989–1999**

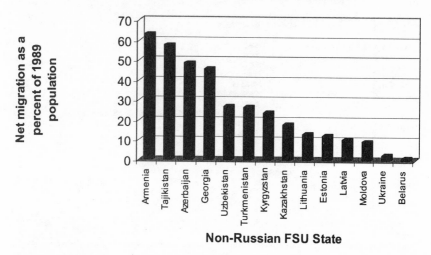

Source: Goskomstat Rossii, *Chislennost' i migratsiia naseleniia Rossiyskoi Federatsii* (various years).

nians of the 14 non-Russian titular nationalities of the successor states have had net immigration to Russia every year from 1989 to 1999. Since 1994, a clear pattern of migration of the 14 non-Russian titular nationalities of the successor states has emerged, as every one has had net immigration to Russia. This is a pattern similar to that experienced by other empires as they broke apart such as Algerians following the French to France and Indians following the British back to Britain. This should not be surprising given Russia's GDP per capita and the size of its economy.

Since the breakup of the Soviet Union, the formerly internal borders among the successor states have been rather porous, as the newly independent states struggle to erect the necessary border control mechanisms and institutions. Figures for the number of illegal migrants in Russia range from 700,000 up to a rather implausible 15 million. One source citing the head of the Russian Goskomstat put the number of illegal migrants at 3 to 4 million. Regardless of the exact number, they seem to indicate that Russia is becoming a migration magnet within the region and in a manner not too different from other migration-magnet countries, which try encouraging legal migration of certain groups while discouraging that of others, and failing at this balancing act. Witness the fact the population count in the 2000 United States census exceeded estimates by 5 to 7 million, largely due to the 8.5 million undocumented immigrants.[27] Anecdotal information indicates that most of these people enter Russia legally and then overstay their visas and most are from

Figure 11.7 **Russia: Net Migration and Natural Increase, 1980–2001**

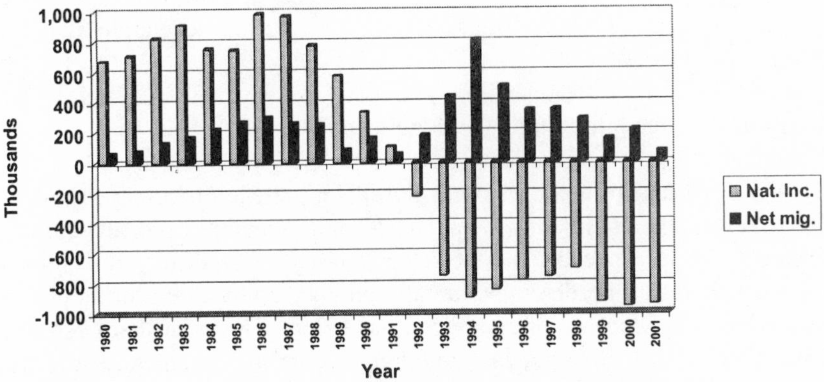

Source: Goskomstat Rossii, *Demograficheskii ezhegodnik* (various years).

the other FSU states as well as large numbers from Southeast Asia. One rather contradictory statement read "according to official data, at the end of 2001 in Russia, there were 5 million illegal migrants." If true, then the population decline may not actually be taking place. There will be a separate effort in the October 2002 census to enumerate these people and obtain information about socio-demographic characteristics and motives for migration. While the extent of criminality often attributed to groups of illegal migrants in Russia may be overstated, it is likely that they are living on the fringes of Russian society. If it is the case that many are from outside the former Soviet Union and lack Russian-language skills, then their incorporation into Russian society may be more problematic. However, that hardly appears to be what Russia desires.

In conclusion: The Russian population stood at 144.0 million on January 1, 2002, down 4.3 million from its peak at the beginning of 1992. Deaths have exceeded births over that period by 7.7 million, with net migration of 3.6 million making up less than half the natural decrease (see Figure 11.7). Though the "mortality crisis" in Russia has received the greatest attention, it is actually the decline in the birthrate that has had the greatest impact on population decline. Since 1987, the annual number of deaths has increased by 720,000 while the number of births has declined by 1,191,000. The trends of extremely negative natural decrease (the excess of deaths over births) and slowing migration that fails to compensate appear to have increased since 1998. In each of the last three years, the natural decrease of the population was over 900,000. In 2001, net immigration into Russia only compensated for 8 percent of this decline, and the population fell by

820,000. As noted above, the return of Russians and Russian-speakers seems to be waning, and thus these trends of increasing negative natural increase and declining immigration will likely conspire to further accelerate Russia's population decline.

Russian Demographic Policy Under Putin

Though Russian President Vladimir Putin is a protégé of former President Boris Yeltsin, the images of personal health and fitness they project couldn't be further apart, especially that of Yeltsin's last years in office. The chronically drunk Yelstin needed to be literally propped up by aides during public appearances in which he often seemed incoherent. This is in sharp contrast to the vibrant, athletic Putin, who if nothing else projects a radically different public face for his beleaguered country. The differences in approaches to the demographic ills of the country are similar. While Yeltsin was barely aware of his own surroundings, let alone his nation's dismal health, Putin, in his first state of the nation address, listed the country's demographic ills as one of the nation's most urgent problems.[28] In subsequent statements and policy initiatives he has discussed the populace's lack of exercise and general poor health, the declining birthrate, excessive alcohol consumption, high smoking rates, the growing number of homeless children, and measures to encourage immigration. He has also placed great importance on Russia's first post-Soviet population census to be conducted in October 2002 to be used as basis for the development of demographic and social policy in the country.[29]

Russia's demographic policy is contained in a document entitled "Conception of Demographic Policy of Russia for the Period to 2015"[30] and various other pieces of legislation. Taken together, these legislative initiatives and public policy pronouncements by Putin and others indicate that the country's demographic future is of considerable concern. The problem that seems to be mentioned most often is simply the dwindling numbers of Russians. Two areas often mentioned to address the problem of the numeric decline in the population are to stimulate the birthrate and encourage further migration to Russia. A $150-million package of measures to encourage women to have more children was introduced to stimulate the birthrate by increasing child allowances and implementing housing credit programs for young families when they have children, as well as other measures designed to improve reproductive health. A law recently passed in the Duma restricts smoking in various public places and on transport, including all Aeroflot flights.

Russia seems to be pinning its hopes on increasing migration as a way to compensate for population decline. However, according to a recent UN study of "replacement migration," in order for Russia to maintain the same population size between now and 2050, there would have to be a net migration of 27.9 million persons, and to maintain a constant labor size there would have to be a migration of 35.8 million persons.[31] Both figures are greater than the size of the Russian diaspora population, the major source of migrants that Russia would like to tap into. Given that there has been a dramatic slowdown in migration from over 800,000 in 1994 to just 72,300 in 2001, it does not seem likely that migration will be a solution to Russia's population decline. Russian President Vladimir Putin and others often talk of stemming Russia's population decline on increased migration into Russia—more specifically, legal, regulated migration of well-educated Russians and Russian-speakers from the non-Russian FSU states to regions in Russia where they are most needed. He has directed groups within the Federal Migration Service to develop a migration policy, and several working groups have been established to develop proposals on regulating migration. Putin believes that the influx of Russians would be a godsend because they all speak Russian, but that these new migrants must be sent to labor-deficit areas where they are most needed. One proposal talked of setting up immigrant camps near Moscow and St. Petersburg where these new immigrants would live for six months. He also wants to crack down on illegal migration. Thus, he seems to want to open the doors to Russians and Russian-speakers, keep out undesirable migrants (i.e., less-educated, Central Asians, Afghans, Chinese) and direct those who do return to depressed areas. The proposed policy is flawed in several respects. As migration-magnet countries elsewhere have proven, it is impossible to completely regulate migration to allow in only selected groups. A country can partiality accomplish this, but full enforcement of migration policy, especially in a country with as long a border as Russia's, is extremely expensive and difficult. In theory, freedom of movement within Russia is guaranteed under the 1993 constitution. While certain regions and cities at times have been able to erect entry barriers, the country has liberalized too much to be able to completely revert to a system where the state could completely direct internal migration movements. But as in many aspects of Putin's presidency he seems to trying to use an element of force to implement policy and has abolished the migration service and placed it under the Interior Ministry, which runs the nation's police service. Putin does not seem to have considered the impact on neighboring states of the potentially destabilizing impact of his idea of trying to lure the remaining Russian diaspora population. Evidence of the difficulty of migration enforcement is the new citizenship law passed by the Duma, which was signed into law in the summer of 2002. The law raises the waiting period for citizenship to five years and requires proof of financial support and Russian-language ability. The Russian

Figure 11.8 **Russia: Historical and Projected Population, 1960–2050**

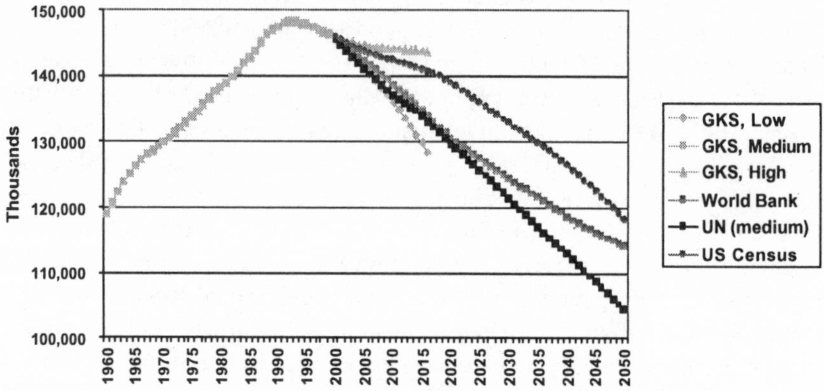

Source: Goskomstat Rossii, *Predpolozhitel'naia chislennost' naseleniia RF do 2016 goda*; World Bank, HNP website; UN, *World Population Prospects: The 2000 Revision*; US Census Bureau, International Database.

diaspora does not receive any special treatment, although the waiting period for former Soviet citizens is only one year. Enactment of the law is expected to greatly slow the influx of immigrants to Russia, both legal and illegal.

It is the levels of fertility and how much they deviate from replacement level (a TFR of about 2.1 children per woman) that have the largest impact on population growth. In Russia, even if the fertility rate were to return to replacement, the overall population would continue to decline, because starting in 2007, the smaller cohorts of women born starting in 1987 will start entering their childbearing years. If Russia's demographic policy were focused on simply increasing the number of people in the country, this would be most easily accomplished through increases in the fertility rate. Efforts to increase life expectancy would have a lesser numeric impact, though they may have a significant impact on the health, quality, and productivity of the population. According to the World Bank projections, the TFR in Russia is projected to rise toward replacement level in the year 2050. The UN figures project an increase to a TFR of 1.75 in 2050.[32] Goskomstat, in its projections to 2015, does not project any sort of increase in the TFR and expects it to remain at a very low level of 1.16 children per women.

For mortality, Goskomstat projects an increase to 69.6 years for both sexes from the current 64.8 years. The level for males is expected to reach 63.9 years, a level last achieved during the anti-alcohol period of the late 1980s. Aside from that somewhat artificial period, the last time Russian male life expectancy was at that level was the early 1960s. Goskomstat expects the world's largest male-female gap in life expectancy to persist well into the future, the difference being about 12 years in 2015. The World Bank projects a somewhat lower life expectancy by 2015 than Goskomstat, 67.5 years versus 69.6 years, but then expects life expectancy for both sexes to reach 75.7 in 2050. Life expectancy for both sexes combined has only been above 70 for a brief period during the anti-alcoholic campaign of the late 1980s.

Based upon these assumptions regarding future fertility and mortality trends, combined with the age structure at the beginning of the projection period, the various projections call for different population sizes in the future, although all project steep population declines. The differences between the three Goskomstat scenarios are levels of projected migration, ranging from zero net migration by 2015 in the low scenario to a positive 122,000 in the high scenario (see Figure 11.8). Based upon these, the projected population in 2015 ranges from 128.3 4 million to 143.7 million. The U.S. Census Bureau projects a population of 118 million in 2050, the World Bank 114 million, and the UN 104 million. Most of the projections were done before the recent steep rise in HIV/AIDS and do not fully incorporate its potentially crippling impact.

Conclusion

For most of the Soviet Union's existence, it was the third most populous country in the world, a feature it inherited from the Russian Empire (see Table 11.1). Its population size was slightly larger than its Cold War adversary, the United States. When the Soviet Union broke up, Russia became the world's sixth largest country and has now fallen to seventh behind Pakistan. According to the U.S. Census Bureau's projections, Russia is expected to fall to ninth in the world by 2025 and further to fourteenth by 2050. Even without incorporating the potentially devastating impact of AIDS, others project Russia's population to be barely above 100 million by mid-century, possibly dropping Russia out of the ranks of the world's twenty most populous nations. Regardless of Russia's exact future population size and world ranking, it is obvious that Russia will fall from the ranks of the world's largest countries, a position that it and the Soviet Union held for much of the twentieth century and upon which it based some of its superpower status. The question is, what impact will Russia's reduced population size have on its ability to influence world politics or even events within the region? Population size is just one of many factors that determine a country's capabilities, others including size of the economy and territory, technological level, size and capacity of armed forces, organizational effectiveness, and morale.[33] Assuming that there are no boundary changes over the next half-century (a

Table 11.1

Russia: World Population Rank, 1900–2050

1900 Rank	Country	Population (millions)	1950 Rank	Country	Population (millions)	1989 Rank	Country	Population (millions)	2000 Rank	Country	Population (millions)	2050 Rank	Country	Population (millions)
1	China	420	1	China	563	1	China	1,123	1	China	1,262	1	India	1,620
2	India	300	2	India	370	2	India	834	2	India	1,014	2	China	1,470
3	Russian Empire	135	3	Soviet Union	179	3	Soviet Union	286	3	United States	276	3	United States	404
4	United States	75	4	United States	152	4	United States	247	4	Indonesia	225	4	Indonesia	338
5	Russia	68	5	Russia	101	5	Russia	147	5	Brazil	173	5	Nigeria	304
									6	Russia	146	6	Pakistan	268
									7	Pakistan	142	7	Brazil	207
												8	Bangladesh	205
												9	Ethiopia	188
												10	Congo	182
												11	Phillipines	154
												12	Mexico	153
												13	Vietnam	119
												14	Russia	118
												15	Egypt	113

Source: U.S. Census Bureau, International Database.

heroic assumption given the events of the past fifteen years), Russia will remain territorially the world's largest country.

There is ample evidence of the deteriorating physical and human conditions of Russia's armed forces. The worsened health conditions of its population and the exodus of some of the country's best and brightest scientists, engineers, and other educated elite do not bode well for the country's continued ability to maintain its technological superiority. Russia's negative demographic trends in terms of both declining quantity and quality are shaping and limiting the country's policy options. Most analysts agree that Russia should simply adjust its expectations to having a smaller population size in the future and use this demographic window of opportunity to focus on improving the health, educational, and other social capital aspects of its population rather than on trying in vain to revise quantitative indicators.[34]

Notes

1. Sergei V. Zaharov and Elena I. Ivanova, "Fertility Decline and Recent Changes in Russia: On the Threshold of the Second Demographic Transition," in *Russia's Demographic Crisis*, ed. Julie DaVanzo (Santa Monica: Rand Corporation, 1996).

2. A. Avdeev, "The Extent of the Fertility Decline in Russia: Is the One-Child Family Here to Stay?" Paper presented at the IUSSP Seminar on "International Perspectives on Low Fertility: Trends, Theories, and Policies," Tokyo, Japan, March 21–23, 2001.

3. Andrei A. Popov and Henry P. David, "Russian Federation and USSR Successor States," in *From Abortion to Contraception: A Resource to Public Policies and Reproductive Behavior in Central and Eastern Europe from 1917 to the Present*, ed. Henry P. David (Westport, CT and London: Greenwood Press, 1999), pp. 223–277.

4. Population Reference Bureau, *2001 World Population Data Sheet*. Available at http://www.prb.org/.

5. Anatoly G. Vishnevskii, "Family, Fertility, and Demographic Dynamics in Russia: Analysis and Forecast," in *Russia's Demographic Crisis*, p. 16.

6. Population Reference Bureau, *2001 World Population Data Sheet*. Available at http://www.prb.org/.

7. Zaharov and Ivanova, "Fertility Decline and Recent Changes in Russia," p. 13.

8. The package of pro-natalist measures consisted of extending the period of maternity leave from one to three years and changing the definition of a large family to three children, making such families eligible for a variety for public services and housing benefits. Ibid., p. 12.

9. John Bongaarts, *The End of the Fertility Transition in the Developed World*, Policy Research Division Working Paper No. 152 (New York: The Population Council, 2001).

10. John Bongaarts, "The Fertility-Inhibiting Effects of the Intermediate Variables," *Studies in Family Planning*, vol. 13, issue 6/7 (June–July 1982), pp. 179–189.

11. Popov and David, "Russian Federation and USSR Successor States," pp. 248–249.

12. These include under-registration of abortions in new, private heath clinics,

incomplete registration of early abortions, and changes in 1991 to the registration system. See Andrei A. Popov, "Family Planning and Induced Abortion in Post-Soviet Russia of the Early 1990s: Unmet Needs in Information Supply," in *Russia's Demographic Crisis*, p. 9.

13. UNICEF, TransMONEE Database, Innocenti Research Centre, Florence, Italy. Available at http://www.unicef-icdc.org/.

14. Julie DaVanzo and Clifford Grammich, *Dire Demographics: Population Trends in the Russian Federation* (Santa Monica, CA: Rand Corporation, 2001), p. 34.

15. UNICEF, TransMONEE Database, Innocenti Research Centre, Florence, Italy. Available at http://www.unicef-icdc.org/; and Megan Twohey, "1 in 3 Babies Born to Unmarried Moms," *The Moscow Times*, November 29, 2001.

16. *Demograficheskii ezhegodnik Rossii* (Moscow: Goskomstat Rossii, 2001), p. 149.

17. UNICEF, TransMONEE Database, Innocenti Research Centre, Florence, Italy. Available at http://www.unicef-icdc.org/.

18. RIA News Agency, "Russia: Birth rate Down, Death Rate Up, Smoking and Drinking to Blame," October 24, 2000, reprinted in Johnson's Russian List, October 25, 2000. Available at http://www.cdi.org/russia/johnson/.

19. Anatoly Vinokur, Joana Godinho, Christopher Dye, and Nico Nagelkerke, *The TB and HIV/AIDS Epidemics in the Russian Federation*, Technical Paper No. 510 (Washington, DC: The World Bank, 2001), p. 29.

20. Christof Ruhl, Vadim Pokrovsky, and Viatchslav Vinogradov, *The Economic Consequences of HIV in Russia* (Washington: DC: The World Bank). Available at http://www.worldbank.org.ru/eng/group/hiv/.

21. UNICEF, TransMONEE Database, Innocenti Research Centre, Florence, Italy. Available at http://www.unicef-icdc.org/.

22. *Demograficheskii ezhegodnik Rossii*, p. 215.

23. Jose Luis Bobadilla, Christine A. Costello, and Faith Mitchell, eds., *Premature Death in the New Independent States* (Washington, DC: National Academy Press, 1997); and Giovanni Andrea Cornia and Renato Paniccia, eds., *The Mortality Crisis in Transitional Economies* (New York and London: Oxford University Press, 2000).

24. Elizabeth Brainderd and Elena A. Vravikova, "Death and the Market," paper presented at the World Bank, Washington, DC, March 13, 2002, pp. 44–45.

25. Shkolnikov and Vallin, "Recent Trends in Life Expectancy and Causes of Death in Russia, 1970–1993," in *Premature Death in the New Independent States*, pp. 34–63.

26. Victor Nesterovich, "Russian Doctors Attack Tobacco," *RIA Novosti*, February 19, 2002.

27. Jeffrey Passel, "New Estimates of the Undocumented Population in the United States," *Migration Information Source*, Migration Policy Institute. Available at http://www.migrationinformation.org/).

28. Murray Feshbach, "Russia's Population Meltdown," *The Wilson Quarterly*, vol. 25, no. 1 (Winter 2001), pp. 15–21.

29. "Census a Priority in 2002, Putin Says," *Interfax Statistical Report* for April 27–May 3, 2002. Available at http://www.interfax-news.com/.

30. "Russia's Population Explosion," *Jane's Information Group Limited*, September 6, 2001, reprinted in Johnson's Russian List, September 7, 2002. Available at http://www.cdi.org/russia/johnson/.

31. *Replacement Migration: Is It a Solution to Declining and Aging Populations?* (New York: United Nations, 2001), p. 27.

32. *World Population Prospects: The 2000 Revision, Vol. 1: Comprehensive Tables* (New York: United Nations, 2001), pp. 388–389.

33. Geoffrey McNicoll, "Population Weights in the International Order," *Population and Development Review*, vol. 25, no. 3 (September 1999), p. 416.

34. DaVanzo and Grammich, *Dire Demographics*, pp. xv, 81.

For Further Reading

Timothy Heleniak, *Migration from The Russian North During the Transition Period*, The World Bank, Social Protection Discussion Paper No. 9925, September 1999.

Timothy Heleniak, "Migration and Restructuring in Post-Soviet Russia," *Demokratizatsiya: The Journal of Post-Soviet Democratization*, vol. 9, no. 4 (Fall 2001), pp. 531–549.

DaVanzo, Julie, and Clifford Grammich, *Dire Demographics: Population Trends in the Russian Federation*, Population Matters, A RAND Program of Policy-Relevant Communication, Santa Monica, CA: 2001.

Goskomstat Rossii, 2002 Russian Population Census website http://www.perepis2002.ru/.

G. Vishnevskii, ed., *Naselenie Rossii 20—ezhegodnyi demograficheskii doklad* (Moscow: Center for Demography and Ecology, Institute for Demo-Economic Forecasting of the Russian Academy of Sciences, annual report).

Demoscope Weekly, electronic version of the bulletin *Population and Society* issued by the Center for Demography and Ecology, Institute for Demo-Economic Forecasting of the Russian Academy of Sciences (http://demoscope.ru/).

Goskomstat Rossii, *Demograficheskii ezhegodnik Rossii: Statisticheskii* (Moscow: Goskomstat, annual publication).

State Committee on Statistics of the Russian Federation (http://www.gks.ru/).

12

The Challenge of Rural Revival

Stephen K. Wegren,
Vladimir R. Belen'kiy, and
Valeri V. Patsiorkovski

When Boris Yeltsin resigned as president of Russia at the end of December 1999, the agricultural sector had experienced almost a decade of steep decline. Despite progress during the 1990s in terms of privatizing agricultural enterprises (both food producing and food processing), as well as the privatization of agricultural land, food production declined precipitously, leading many analysts to conclude that agrarian reform during the Yeltsin years had not yielded successful results. One could easily argue, based on production data alone, that the agricultural sector was worse off at the end of the decade than at the beginning. Indeed, it is hard to imagine how the agricultural situation could have been more challenging. When Vladimir Putin assumed the post of acting president in January 2000 (Putin was elected president in his own right in March 2000), the list of rural problems he inherited was long and complex:

- declines in animal stocks during the Yeltsin years exceeded those of Stalin's collectivization, and by 2000 were at the levels of the 1950s.
- total food output was down an estimated 40–50 percent (in ruble value) compared to 1990 levels.
- food imports, particularly of meat and poultry, increased substantially during the 1990s.
- crop yields declined and were at their lowest levels in decades.
- farm efficiency was worse than during the late Soviet period, and farm unprofitability soared.
- rural capital investment declined significantly, and rural infrastructure was collapsing.
- per capita food consumption declined for proteins and most carbohydrates, increasing only for starchy foods.

- the rural demographic situation worsened.
- the amount of agricultural land under cultivation declined, and land reclamation virtually ceased.
- in the latter 1990s, food aid was provided by the West in order to prevent starvation in some regions of Russia.
- the rural sector in general suffered from demodernization, as fewer mineral fertilizers were applied, labor became more manual, and farms' capital stock had become depleted.

A key aspect to Russia's economic recovery is the revival of its agricultural economy. Not only does agriculture exert a significant economic influence, despite declines in its contribution to GDP, but it is inconceivable that a market economy will develop without the participation of rural economy. Rural revival has broader implications for the further development and consolidation of market reforms. Likewise, rural revival is important for the consolidation of political reforms, as rural elites and the population at large are needed to support democracy. In short, the fact that 28 percent of the Russian population lives in rural areas means that rural Russia has importance for the nation's future. The purpose of this chapter is to analyze the challenges that Putin faces in trying to revive Russia's rural sector. The chapter is divided into three main sections. The first section examines the socioeconomic challenges of rural revival. The second section analyzes the challenges associated with rural land privatization. The third section assesses the prospects for successful rural revival.

Socioeconomic Challenges

In order to understand the challenges that Putin faces in reviving the rural economy, a review of trends is necessary. The socioeconomic challenges of rural revival may be divided into several subsections. Each of these challenges is examined in turn.

Food Production and Consumption

To start, we should note that reference to the "rural economy," "agricultural sector," or "food producers" implies that there is one major food producer in Russia. That is not exactly correct. During the Soviet period, food production came from two main sources: state and collective farms, which produced about 75 percent of the nation's food, and family subsidiary plots (often referred to as private plots), which accounted for approximately 25 percent of the nation's foodstuffs, primarily meat, milk, eggs, and wool. In post-Soviet Russia, the number of producers has changed, as has the structure of output. With the creation of a family farming stratum based upon private ownership and use of land, a third food producer was born. During the 1990s, private farms accounted for about 2 percent of total food production, although the exact percentage varied by type of product. Private farms tended not to engage in much animal husbandry, while concentrating on more profitable crops such as grains, sunflower seeds, and sugar beets.[1]

Beyond the modest contribution of private farmers, the structure of food output changed. In general, food output from large agricultural enterprises (a term that refers to former state and collective farms that had been privatized, as well as to collective farms that retained their status), declined. In 1992, large agricultural enterprises produced 67 percent of the nation's food (measured in ruble value), a sum that declined to 43 percent in the year 2000.[2]

The third food producer is the Russian household. Obviously, households may be located in either urban or rural areas. Of particular interest is the rural household, which tends to produce food for consumption and for sale, while urban households tend to produce food primarily for fun, weekend relaxation, and family consumption. Food output from all households increased during the 1990s, accounting for 32 percent of the nation's food in 1992 and 54 percent in 2000. Regarding this increase, however, three points should be noted. First, there is great regional variation, as would be expected in a country the size of Russia. Some regions experienced significant increases in household production, some modest, and some hardly at all.[3] Second, most of the volume increase occurred during 1992–1995, with household production leveling off after 1995. Third, most of the percentage increase in the nation's food supply was the result of declining output from large farms. Large farm production actually declined more than household production increased. This often-ignored fact is indicated by data that show that in terms of physical output, large farms were producing in 2000 at about 39 percent of their 1990 level, while households were producing at 121 percent of 1990 levels.[4] In short, large farms' production declined some 61 percent, while households' production increased 21 percent.

During the 1990s, the protective environment that had sheltered state and collective farms during the Soviet period disappeared. During the reform period, policy choices were made, and large farms and nascent private farms bore the brunt of the consequences from those policies. Households, which operated small land plots (usually up to 1 hectare in size, although some were larger), did not feel the effects as strongly and the discussion below does not apply to them. Large farms and to a lesser extent private farms were exposed to wholesale price discrimination for their products, to an end to or drastic curtailment of production subsidies, to foreign competition from

Table 12.1

Mean Annual Agricultural Production, 1976–2000 (all categories of farms)

	1976–1980	1981–1985	1986–1990	1991–1995	1996–2000	1996–2000 as % of 1976–1980
Grains (mil. tons)	106.0	92.0	104.3	87.9	65.2	61
Sugar beets (mil. tons)	25.4	25.1	33.2	21.7	14.04	55
Sunflower seeds (mil. tons)	3.1	2.9	4.1	3.8	3.3	106
Potatoes (mil. tons)	40.9	38.4	35.9	36.8	34.4	84
Vegetables (mil. tons)	10.4	12.1	11.2	10.2	11.4	110
Meat and poultry (mil. tons, dead weight)	8.31	9.45	11.41	8.82	4.72	57
Milk (mil. tons)	48.2	48.7	54.2	45.4	33.5	69.5
Eggs (billion)	36.7	43.1	47.9	40.3	32.7	89

Numbers have been rounded.
Grain totals after cleaning.

Sources: Rossiiskii statisticheskii ezhegodnik 2001 (Moscow: Goskomstat, 2001), pp. 407, 416; and authors' calculations.

foreign exports and foreign processed foods, and to enormous increases in the cost of production as inputs, fuels, farm equipment, and machine parts all became more expensive. In short, the economic environment turned decidedly hostile to large farms.

The consequence of a hostile economic environment was a reduction in the amount of food produced. Large farms in particular reduced livestock herds and the amount of land under cultivation, while private farms and households were increasing possession and use of both. Overall, food production declined in absolute terms and relative to production levels achieved in the recent past. Food production trends during 1976–2000 are illustrated in Table 12.1.

The table shows that the 1996–2000 period had the lowest annual mean production levels for all types of products with the exception of vegetables, a staple grown by households. Thus, the supply of domestic food was lower during the second half of the 1990s than during any previous period dating to the 1970s.

The reduction in the supply of domestically produced food was due, in part, to reduced demand and changing consumption patterns by Russian consumers. During the 1990s, Russians consumed less protein (meat), milk, and fatty products, while increasing their intake of cheaper products such as starches (potatoes), vegetables, and carbohydrates (grain products, including pastas and bread products).[5] There were regional variations in consumption patterns, of course, but in general most Russians ate less and everyone experienced a change in their diet. For example, from 1990 through 1999 the percentage of caloric intake from grain products increased from 32 to almost 41 percent, milk and milk products decreased from 17.1 to 9.8 percent, and meat and meat products declined from 15 to 11.8 percent.[6]

In addition, family size had a significant influence on consumption patterns. A straight line descent in the number of kilograms consumed was evident as family size increased (in short, an inverse relationship existed between quantity of consumption and family size). The biggest differences were found for high-preference items such as meat and meat products, as families with four or more children ate about one-half the number of kilograms as did families with one child. Lower preference, cheaper products such as grain products and potatoes exhibited smaller differences in consumption between large and small families.[7]

Foreign Imports and Food Security

The reduction in the supply of domestically produced food was partially compensated by foreign imports. During the Yeltsin years, the composition of Russian imports changed from their previous patterns.[8] The change was due in large part to the significant decline in animal herds, which reduced demand for feed grains and thus the need for grain imports. As a consequence, Russia imported much less grain in the 1990s than it had during the 1980s.[9] Measured in dollar value, meat imports became the most important food import.

Meat and poultry imports increased during 1992–1997, and in 1997 accounted for 30 percent of domestic supplies. Following the collapse of the ruble in August 1998, meat and poultry imports declined significantly in the third and fourth quarters of 1998, and overall fell to 25 percent of total supplies for 1998. The downward trend continued, and in 1999, meat and poultry imports decreased as a percentage of total supplies (22 percent) and in absolute volume, even though imports increased significantly from the U.S. and the European Union (in part due to food assistance following the economic crisis of August 1998). In 2000, meat imports continued to decline due to import restrictions and a shift in trade patterns, with much less food trade being conducted with nations of the far abroad. In 2000, meat and poultry imports accounted for 21 percent of the total national supply.[10]

The importance of food imports during the 1990s is seen by the amount spent. Russia spent more on food imports than was allocated to the agricultural sector in the federal budget (although the consolidated federal budget accounted for only about one-half of total state expenditures). The sum of imports as a percentage of total food supplies reached its peak in 1997. In that year, Russia spent more than 13 billion dollars on food imports. Thereafter, the sum spent on foreign imports declined, falling to 7.4 billion dollars in 2000.[11] By way of contrast, in 2001, which witnessed a large increase in federal funds allocated to agriculture, less than 1 billion dollars from the federal budget was assigned to support domestic producers.[12]

The increase in imported food had political and economic consequences. Politically, since the mid-1990s there has been growing advocacy for Russian "food security," and this movement has gained strength over time. Starting first with agrarian conservatives, concerns over food security spread to agricultural academic institutes, the printed media, and finally to politicians. These voices criticized increasing imports and the relatively open foreign trade policy pursued by Russia since market reforms began. Advocates of national food security support higher trade barriers in the form of tariffs or even import quotas in order to protect domestic producers. It would be fair to conclude that "food security" has become part of the everyday political dialogue among policy makers and is considered a cornerstone of the nation's food policy.[13] Food security is equated with overall economic security, which in turn is a key component of national security. For example, Aleksei Gordeev, now Minister of Agriculture, argued prior to becoming Minister that "food security of the Russian Federation is a component part of its economic security."[14] In June 2000, the government's draft of "priority tasks" in social-economic policy was published. This document indicated in the section on the agro-industrial complex that the nation's food security was a priority goal of the Putin administration.[15] During the following two years, Russian agriculture became increasingly protectionist, for example by placing quotas on sugar imports. In May 2002 Gordeev announced that import quotas would be placed on all meat products.

Ironically, appeals for national food security—which had its origins in poverty and concerns over the percentage of caloric intake coming from imported foods—have not abated even as domestic food production rebounded during 1999, 2000, and 2001. Gross agricultural output grew during those three years for the first time since reforms were begun (see below). However, judging from the frequency of articles in the agricultural press, the push for national food security has not slackened. In March 2002, a conference sponsored by the Ministry of Agriculture was held in Moscow on "Russian Food Security." Two days prior to the opening of the conference, Russia imposed a ban on chicken imports from the United States, in what many perceived as retaliation for U.S. tariffs on Russian steel exports to the U.S. The single largest exporter of poultry to Russia is the United States. The ban on chicken imports lasted about one month and was partially lifted in mid-April 2002, although there were reports that Mayor Luzhkov in Moscow was resisting U.S. imports in late May 2002.[16] The Russian side denied linkage between the two trade issues and instead maintained that the ban was imposed due to concerns over health standards.[17] This was the second ban on U.S. chickens between the two sides, the first occurring in 1996. It is clear that denial of the Russian market to U.S. exporters is one of the weapons in these mini-trade wars. This is important because as the Russian economy has rebounded under Putin, the Russian demand for imported meat is perceived to be quite lucrative.[18] In Russia's largest cities, it is estimated that no less than 70 percent of meat supplies comes from imports.

Beyond the nuisances that protectionist practices and mini-trade wars cause in bilateral U.S.-Russian relations, the potential economic consequences of Russian food security are more severe. President Putin has two conflicting policy goals: to provide national food security by limiting food imports, and to join the World Trade Organization (WTO)—an international organization that is based upon relatively free global trade—perhaps as early as 2003 or 2004 if Putin gets his way. In March 2001, Russia indicated it was committed to joining the WTO, but would need up to seven years after entry to comply fully with all requirements.[19] Whether or not this position is acceptable and how fast negotiations conclude are future questions that will need to be resolved. Putin's address to the nation in mid-April 2002 again stressed the strategic and economic benefits that would accrue from WTO membership. It remains to be seen whether Russia will be given concessionary terms, as was China, and therefore these two policy goals are incompatible.

Financial Challenges

During the 1990s, two broad financial trends were evident which greatly affected the performance of the agricultural economy: price scissors, which led to increased farm enterprise debt; and reductions in capital investment, which contributed to the deterioration in rural infrastructure. Each will be discussed in turn.

"Price scissors" refers to the relationship between the cost of inputs and services relative to the price received by farms for their production. The "scissors" represents a relationship in which input prices are increasing much faster than agricultural prices. In short, farm incomes are declining in real terms. Estimates by Russian experts suggest that during the 1990s, input prices increased four to five times faster than did prices received by food producers, with the greatest gaps occurring at the beginning of the 1990s. Price relationships during the 1990s are illustrated in Table 12.2.

With the onset of reforms, the data show that agricultural prices lagged

Table 12.2

Relationship of Agricultural Prices to Industrial Prices and General Inflation Rate, 1992–2000 (in percent)

	1992	1993	1994	1995	1996	1997	1998	1999	2000
Increase in general consumer prices	2,600	940	320	230	21.8	11.0	84.4	36.5	20.2
Increase in prices for industrial goods	3,380	990	330	270	25.6	7.5	23.2	67.3	31.6
Increase in prices for agricultural products	940	840	300	330	43.5	9.1	11.1	99.8	36.5

Sources: Rossiia v tsifrakh (Moscow: Goskomstat, 1995), p. 187; *Rossiia v tsifrakh* (Moscow: Goskomstat, 2001), p. 332.

behind the general inflation rate in five of the nine years; prices for industrial prices exceeded the general inflation rate in seven of the nine years. A main consequence of lagging agricultural prices was increased farm debt. According to official data, by 2000, more than 89 percent of agricultural enterprises had overdue debt.[20] In early 2001, total farm debt was estimated at more than 229 billion rubles, of which 150 billion rubles was overdue debt, including debt to suppliers, to creditors (including the state and banks), and to farm workers.[21] During 2001, farm debt continued to increase, exceeding the level of farm profit achieved in 2000 by a factor of 10.[22]

In turn, increased farm debt had secondary consequences, including wage arrears, payment of wages in kind, the proliferation of barter between agricultural enterprises and input suppliers. In 2000, nearly three-quarters of farm enterprises were behind in wage payments to their workers.[23] Poor financial conditions also led to the curtailment of production capacity by reducing animal stocks and reducing the amount of land under cultivation. Price scissors led to an increase in the percentage of unprofitable farms, reaching its peak in 1998, with 89 percent of large farm enterprises being unprofitable. As agricultural prices improved relative to production costs during 1999 and thereafter, the number of unprofitable farms declined to 54 percent in 2000 and 52 percent in 2001.

The second broad financial challenge is rural capital investments. During the 1990s, rural capital investments from the federal budget declined. Rural capital investments as a percentage of all federal investment monies declined from 18 percent of the national budget in 1991, continuing until it bottomed out at 2.5 percent in 1997.[24] By the late 1990s, per capita investments into the rural economy were one-fifth the level in urban areas.[25]

The consequences of declining federal investments were twofold. First, farm enterprises assumed a greater portion of responsibility for rural capital investments. In 1994, the "basic source of financing construction projects [was] the resources of the agricultural enterprise itself."[26] In 2000, it was reported that farms were financing 74 percent of rural capital investments.[27] Second, as federal investment monies for rural infrastructure declined, greater responsibility for capital investments fell on regional budgets. In the overwhelming number of cases, however, regional administrations were no more capable than farms of funding capital investments for rural infrastructure.

As a result of these consequences, the construction of rural infrastructure became a fraction of previous levels. One scholar has aptly termed the process the "demodernization" of rural Russia.[28] The construction of rural infrastructure was drastically curtailed, as seen by the fact that by the end of 1994 the construction rate of rural dwellings and rural schools had been halved, and the construction of rural roads was one-third compared to its 1991 level.[29]

By 1996, the construction of rural kindergartens had declined by a factor of 14, hospital beds by a factor of five, and clubs and recreational facilities by a factor of eight.[30] The fact was that the rural infrastructure, which never did attain qualitative or quantitative levels of urban Russia, experienced a significant deterioration during the 1990s. By the end of the decade, nearly all urban settlements had running water, but only 25 percent of rural settlements did; 96 percent of urban settlements had sewer systems, but only 4 percent of rural settlements did.[31]

In addition to a deteriorating infrastructure, during the first half of the 1990s a significant decline in rural services occurred. Social services were curtailed; clubs, as well as other service and recreational facilities, were closed. Rural schools fared even worse, as the number of preschool and educational institutions decreased. The number of students attending rural schools fell by about 66 percent during 1992–1996. By the end of the 1990s, an estimated 700,000 children aged 7–15 who lived in rural areas did not attend school at all.[32] It is no wonder that when asked about their family's material condition, only 10 percent of workers on strong agricultural enterprises, 6.5 percent on average farms, and 6 percent on weak farms responded that their situation had improved since reform was started.[33] These financial realities contributed to socioeconomic and demographic problems in the countryside.

Socioeconomic and Demographic Challenges

The financial trends examined above have had two distinct socioeconomic consequences affecting rural demographics. First, during the 1990s, rural incomes (from wages and salaries) became highly differentiated from urban incomes. According to official statistics, in 1990, on average, a collective farm worker received 88 percent of the monthly monetary income of an industrial worker, and a state farm worker received 104.5 percent.[34] By the end of 1994, a worker in agriculture received an average of 49 percent of the monthly income from wages as did an industrial worker, and 50 percent of the national average monthly income.[35] By the mid-1990s, average monthly agricultural incomes from the place of work were the lowest compared to other branches of the economy. Not only did income urban-rural differentials widen, but rural incomes became increasingly demonetarized. As farm debts rose, farm managers were forced to pay salaries with produce in-kind due to a lack of cash. As a result of these trends, in 1998 almost one-half of the rural population had an income below the minimum subsistence level as defined by the Russian government.[36] Although rural standards of living became harder to measure with precision as rural

Table 12.3

Rural Evaluations of Family Material Conditions, 1993–2000 (in percent)

	March 1993	Nov. 1996	Nov. 1997	Nov. 1998	Nov. 1999	Nov. 2000
Very good or good	5.4	5.5	3.4	2.0	4.8	4.3
Average	44.6	34.9	39.7	31.7	38.2	41.8
Poor	36.6	39.5	37.5	40.1	38.5	34.4
Very Poor	11.2	17.0	15.8	24.7	17.2	16.9

Sources: "Nastroeniia, mneniia i otsenki naseleniia," *Monitoring obshchestvennogo mneniia: ekonomicheskie i sotsial'nye peremeny*, no. 3 (May–June 1994), p. 39; ibid., no. 1 (January–February 1997), p. 64; ibid., no. 1 (January–February 1998), p. 72; ibid., no. 1 (January–February 1999), p. 73; ibid., no. 1 (January–February 2000), p. 76; ibid., no. 1 (January–February 2001), p. 82.

incomes became demonetarized,[37] evaluations about rural material conditions are reflected in the survey data illustrated in Table 12.3. These evaluations are important because poor economic prospects and perceptions about a hostile economic environment for rural families influenced the extent to which reform policies were supported.

The second consequence of financial trends affects rural health care and rural demographics. Rural health care, which never was of particularly high quality, became significantly worse during the 1990s. The deterioration of rural health care was characterized by high rates of infant mortality, an increase in the rural death rate, and a decrease in the rural birthrate.

Broad demographic trends are captured by a statistic called the natural increase coefficient (a ratio of births to deaths per 1,000 persons), or simply the population coefficient. This coefficient deteriorated significantly since the introduction of reforms. For Russia as a whole, the rural population coefficient declined from +2.2 in 1990 to −7.2 in 2000.[38] Many regions experienced double-digit declines for most or all of the 1990s. The impact was the following. Starting in 1991 and continuing through 1994, the rural population increased for the first time in several decades, fueled by migration from the near abroad who settled in rural areas. Starting with reforms in 1992, however, the rural population coefficient turned negative for the first time in many decades. Whereas migration into Russia compensated for a deteriorating population coefficient during the early 1990s, commencing in 1995 the rural population continued its decades-long decline. Once migration into rural Russia declined after 1994, the rural population began to contract due to a large negative population coefficient (an average of 208,710 persons per year). In short, without migration into rural areas and changes in administrative borders, the rural population would have shrunk by almost 210,000 persons a year during 1992–2000, the largest declines since the 1960s. Combined with low salaries, a deteriorating infrastructure, poor health, education, and social services, anecdotal evidence suggests that out-migration by rural youth intensified, thereby further weakening the foundation of human capital on which to consolidate agrarian reforms.

The Challenge of Land Privatization

Land privatization is important for Russia's transition for several reasons. The privatization of land creates political barriers against future state incursions against individuals and their property, thereby helping to develop a civil society independent of the state. Economically, privatization creates individual incentives to modernize and increase output. Privatization also creates opportunities for investors. The challenge of land privatization may be divided into two sections, legal and economic. Each will be discussed in turn.

Legal Challenges to Land Privatization

Russia's land relations for much of the twentieth century were governed by the 1922 Land Code. This Land Code forbade the ownership, buying, and selling of land. The 1922 Land Code was in effect until 1991, when it was replaced by another Land Code. The 1991 Land Code liberalized previous restrictions by allowing leasing of land and land ownership. Although the 1991 version permitted the ownership of land, it restricted the purchase and sale of land.

In the post-Soviet period, a new Land Code was needed to replace the Soviet-era code, and to act as a legal instrument for the regulation of land relations in a new political and economic environment. During 1994–1998, debates on a new Land Code occurred in 14 plenary sessions of the State Duma. On three occasions, a Duma-sponsored version was passed, only to be rejected either by the Federation Council or the president.[39] During this time period, the Land Code was one of the most contested and contentious issues in executive-legislative relations. The central issues of disagreement were over the rights of buying and selling land, in particular agricultural land, and whether, once purchased, agricultural land could be converted into urban-use land. Owing to this disagreement and the lack of a Land Code, land relations were regulated by presidential decrees issued in October 1993 and March 1996, as well as by the 1993 Constitution (Article 27) and 1994 Civil Code (Articles 260 and 261), both of which explicitly permit the ownership of land.[40]

In July 1998, a presidential version of the Land Code was submitted to the Duma. This version was discussed, and amendments debated, for more than two years. Finally, a compromise was reached in which it was agreed that major elements of the 1998 presidential draft would be accepted if purchases and uses of agricultural land would be treated separately in a different law. In early 2001, an entirely new version of the Land Code, which excluded agricultural land, was drafted by the government.[41] The Putin government was actually working on more than one front, indicated by the fact that in March 2001, Chapter 17 of the Civil Code was amended (by a vote in the Duma) allowing the buying and selling of nonagricultural land.[42] In April 2001, the new version of the Land Code was signed by Prime Minister Kasianov and submitted to the Duma, passing in its first reading in June 2001 by a vote of 251–22. In July 2001, the draft Land Code passed its second reading by a vote of 253–153.[43] The second reading of a bill is considered critical because it reflects any changes from the first reading, and once passed the second time, successful passage in the third reading is often assured. In late September 2001, the Land Code passed its third reading by a vote of 257–130.[44] In early October, the Federation Council approved the draft 103–29, and forwarded the code to President Putin for his signature. Finally, toward the end of October 2001, Putin signed the Land Code into law, thus ending eight years of controversy over land relations.[45] Russia finally had a legal document that permitted not only land ownership, but also the buying and selling of nonagricultural land.

On the one hand, the new Land Code gives an important legal and psychological boost to the Russian land market. The new Land Code provides a legal framework for land ownership, and specifies the rights of foreigners. According to the new Land Code, owners of buildings have the right to purchase the land on which their buildings stand, whereas previously they could only rent this land. Henceforth, buildings will be sold with the land on which they stand. Owners of buildings located on land owned by a third party will have the right to purchase or lease this land. The new Land Code also defines the rights of foreigners. Foreigners can lease or purchase land. However, for land purchases, foreigners cannot purchase land along Russia's borders or within special zones established by federal law.[46]

On the other hand, there are several shortcomings to be noted in Russia's new Land Code.[47] First, it is estimated that the Land Code will regulate land relations for about two percent of Russia's total land mass, in short, mostly urban land in cities are affected. Second, many articles in the Land Code are already in effect in different laws. Third, several articles in the Land Code refer to legislation not yet adopted and not in force, which makes those parts of the Land Code invalid. Fourth, the Land Code does not address the question of land turnover between enterprises of different juridical forms (joint stock companies, cooperatives, municipal, private, etc). Fifth, certain chapters in the Land Code are overly abstract and vague, or in other cases compliance appears to be voluntary. Finally, certain phrases are unclear as to their meaning. Shortcomings such as these, as well as others, reflect the haste with which the document was drafted and the inexperience of those who drafted the document in land relations.[48]

Economic Challenges to Land Privatization

Prior to the passage of the Land Code in October 2001, the Russian land market was restrained by economic factors. Starting in 1994 and continuing thereafter, millions of land transactions (of all types) occurred annually. In order to understand trends in the Russian land market, it is first necessary to explain that the land market consists of two types of submarkets: the municipal market and the private market. The municipal market involves land transactions from state or municipal land funds; the private market refers to the sale of land between private citizens. In terms of area and number of transactions, land leasing (from both types of submarkets) has been more prevalent than land purchases, accounting for more than 90 percent of land transactions annually during the 1990s. However, as the decade progressed, two trends were evident: (1) the purchase and sale of land became more significant, rising to about 6.5 percent of all land transactions in 2000, as leasing transactions declined somewhat; (2) "market" transactions of land (transactions involving monetary exchange, as opposed to nonmonetary exchanges) also increased. These trends are illustrated in Table 12.4.

It should be noted that the rural land market accounted for about 42 percent of all purchase transactions in 2000 (all types of uses).[49] The number of land purchase transactions in rural regions increased for five consecutive years, through 2000. Despite upward trends, it is likely that the rural land market has been more constrained by economic factors than by legal restraints. This position is reflected in the following characteristics of the agricultural land market.

1. Small sizes of land plots. Even though the prices of rural land plots tend to be low, the average size of a rural land plot in a private transaction (for all types of uses) in 2000 was .19 hectares (190 square meters).[50] With mean monetary incomes of less than $29 per month per rural worker, the purchase of small land plots is hardly surprising.[51]
2. Type of demand. The most popular uses for rural land, purchased from either municipal or private sellers, are small-scale agricultural operations, in particular subsidiary plots or fruit and vegetable gardens.

Table 12.4

Structure of Land Transactions in the Russian Federation, 1998–2000 (all transactions, urban and rural)

Type of transaction	Number of transactions 1998	% of total	Number of transactions 1999	% of total	Number of transactions 2000	% of total
Total land transactions	4,415,524	100	5,226,629	100	5,271,185	100
Leasing of state or municipal land	3,995,491	90.49	4,733,347	90.56	4,728,699	89.7
Sales of lease rights of state or municipal land	—	—	8,104	.16	15,615	.3
Sale of state or municipal land	11,467	.26	22,191	.42	23,971	.45
Sale-purchase of privately owned land	234,590	5.31	291,771	5.58	315,508	6.0
Gift	26,452	.6	22,443	.43	23,441	.44
Inheritance	144,735	3.28	147,533	2.82	161,912	3.07
Mortgage	2,789	.06	1,240	.02	2,039	.04

Source: Gosudarstvennyi (natsional'nyi) doklad o sostoianii i ispol'zovanii zemel' Rossiiskoi Federatsii v 2000 godu (Moscow: Federal Land Cadastre Service of Russia, 2001). p. 111.

3. Purchases where land sales were already liberalized. In regions where land sales were legalized prior to the passage of the Land Code, demand for rural land was low, and often only about one-half of available plots were sold in auction.
4. Sensitivity to amenities and infrastructure. Survey data from the Institute of Land Relations and Land Tenure in Moscow and the Institute of Sociology in Kaluga oblast show an acute sensitivity to the quality of land, where it was located, and what amenities were located on or near the land.[52] For well-situated, good-quality land with amenities, demand greatly outstripped supply. For poor-quality land, located in remote regions, with few or no amenities, demand was minimal and supply exceeded demand. In this respect, underdeveloped or deteriorating rural infrastructure is a key element influencing demand for land.

Prospects for Successful Rural Revival

Previous sections have surveyed and discussed the challenges to be overcome if there is to be a sustained rural revival. This section will examine the prospects for success. Before turning to a discussion of each challenge, it would be useful to review the general orientation of Putin's agrarian policy. Early policy orientations identified the agricultural sector as an important segment of the economy. Immediately after Putin became acting president, there were indicators that the revival of food production was considered one of the strategic policy directions under President Vladimir Putin. In February 2000, Acting President Putin attended a national conference on the problems of the agro-industrial complex in the Krasnodar krai. In his speech to the conference, Putin stated that "Our first-order task is to raise the volumes of food output to the levels they were at the end of the 1980s to the beginning of the 1990s and to appreciably reduce the country's food dependence on imports."[53] Just a few months after Putin was elected as president, a program entitled "Basic Directions of Agrofood Policy to 2010" was presented to the Cabinet by Deputy Prime Minister and Minister of Agriculture Aleksei Gordeev.[54] This new program, introduced in July 2000, posited three broad tasks: (a) to develop and strengthen market conditions in the rural economy; (b) to stabilize food production; and (c) to achieve the first two tasks in the shortest time possible. Thus, after years of indifference during the Yeltsin era, the condition of Russian agriculture once again became important to policy makers, with the goal to create an economic environment supportive of domestic production.

Food Production and Consumption

In recent years, agricultural production as a whole has improved—increasing four percent in 1999, five percent in 2000, and almost seven percent in 2001—marking the first growth in food production since the late 1980s.[55] Within this growth trend, the cultivation of plant products has revived more quickly than for animal husbandry. During 1999–2001, for example, the ruble value output from plant products increased 9, 12.5, and 10.3 percent, respectively, while animal husbandry experienced –.7, .6, and 2.6 percent growth.[56]

In general, the questions about continued increases in domestic food production are twofold: Will increases continue, and which sector (plants or animals) will benefit the most? It is hard to predict production trends since they are dependent upon more than state policy. Weather, drought, and availability and price of fuels affect how much land is cultivated, the price of fertilizers and how much is applied, disease—all are relevant factors. What is clear is that the Putin administration favors domestic production over food imports and has taken several important steps to allow domestic producers to improve their performance. Some of these measures are discussed below. Thus, in general, barring unforeseen natural calamities such as drought or insects, the prospect for continued growth in agricultural production is better than it has been for at least a decade. Turning to the question of which sector is likely to benefit most, it is hard to envision the animal-husbandry sector reviving in the short-term. The losses to cattle and pig stocks were so severe that it will be many years before they recover to the levels of 1990. The number of poultry has rebounded more quickly, but for this commodity domestic producers face heavy pressure from foreign imports. With domestic meat production in such a slump, the government has not been eager to impose import quotas on meat products (with the exception of poultry meat). Finally, as a high preference item (more expensive), demand for meat is closely linked to consumer incomes.

Conversely, production of plant products is able to rebound more quickly, as the 2000 and 2001 harvests showed. Moreover, the production of profitable industrial crops—those used to make oils, such as sunflower seeds and sugar beets—never experienced the sharp decline in cultivated area or output as did food grains. In short, for a variety of reasons, plant production is likely to continue to exhibit more growth.

Food consumption trends are also difficult to predict because they depend not only on production, but even more so on consumer incomes. As real incomes increase, it may be expected that per capita food consumption will also rebound. Already, a mini-trend was evident in this direction during 1999–2000. The improvement in real incomes is reflected in food consumption patterns and conform to the predictions of "Bennett's law," which postulates that the proportion of calories coming from starchy foods decreases as incomes rise.[57] Russia's food consumption data support this theorem, although it remains to be seen how long the trend will continue.[58]

Foreign Imports

A more protectionist food import policy did not originate under Putin, but actually in the latter Yeltsin period. The financial crisis of 1998 gave a huge impetus to higher protection for domestic producers and food security. One of the strongest proponents for more protection for domestic producers was former Minister of Agriculture Viktor Semenov (March 1998–May 1999), who advocated a system of higher tariffs and quotas on imported food products that compete with Russian products, and lower tariffs on products that Russia did not grow.[59]

Under Putin, protectionist tendencies have grown. Putin himself remarked that tariffs on imports should be used not only to protect domestic producers, but also to stimulate production of high-quality products.[60] Minister of Agriculture Gordeev supports the use of custom and tariff policies to protect domestic producers, and this policy was included in the government's ten-year agricultural plan to 2010. In May 2001, Gordeev spoke out for the use of import quotas on foodstuffs, stating that "we need to increase the protectionist role of the state with regard to our food producers and protect them from unfair competition."[61] With tariffs increasing, and more widespread use of quotas, the prospects are that Russia will remain protectionist, with the dilemma it creates for broader trade policy as noted above.

Russia's agricultural trade protection actually has two dimensions. First, although federal policy is more protectionist under Putin, the biggest restraint on domestic and international food trade is at the regional level. Regional leaders are sensitive to food security for the populations in their regions. Thus, a majority of regions have adopted either restrictions on food imports or food exports.

Second, it should be noted that Russia does not face a single trade challenge. Most meat imports into Russia come from the European Union, the U.S., and China (the so-called far abroad). Within the Commonwealth of Independent States (CIS), Russia is a primary importer of grain, most of which comes from Kazakhstan. (Russia also exports grain, most of which is sold to other CIS nations). Within the CIS, Russia has a number of trade advantages, which makes food trade profitable and facilitates overall integration. In mid-2002 Russia and Ukraine signed a bilateral free trade agreement. Beyond bilateral trade relations, Russia faces two trade challenges, each qualitatively different. On the one hand, Russian leaders want to protect domestic producers from meat imports from the far abroad. This policy is necessary to allow domestic animal husbandry to recover and become competitive. On the other hand, within the CIS (the near abroad), Russia wants to see more open markets in the other member states, many of whom are even more protectionist than Russia. International food trade among CIS nations remains restricted for a number of reasons, including undeveloped infrastructure, insufficient transportation, and regional policies that prohibit exports.[62] In sum, international food trade policy is somewhat dichotomous in Russia.

In the short term, at least, this balancing act can be maintained. Russia intends to continue import protection.[63] Somewhat oddly, Russia has lower wages, lower transportation rates, and lower energy costs than its competitors in America and Europe. Nonetheless, Russian agricultural production costs more than in those nations, and also does not benefit from the full range of production and export subsidies as do European Union nations. With energy prices increasing in Russia, it is possible that Russia will be "locked" into protectionist policies.

Some import protection will not be an obstacle if Russia is granted concessionary terms for membership in the WTO—the exact terms will need to be negotiated. The more difficult problem will be entry into the European Union, but that prospect is so far off—at least a decade and most likely even longer—that it is not even relevant at present. Thus, the likelihood is that Russia's food trade policies will remain stable and are unlikely to change significantly in the next few years.

Financial Prospects

In order to revive the financial condition of large agricultural enterprises, which the Putin administration has indicated is the key to overall agricultural revival, three policy initiatives have been launched.[64] First, the Putin government introduced a debt amelioration plan for large agricultural enterprises and private farms. In June 2001, Prime Minister Kasianov signed a government resolution for the restructuring of agricultural debt.[65] Farm debt was divided into several categories: taxes, insurance payments, and penalties and fines. Overall, the plan envisioned a ten-year process: Tax and insurance payment debts would be restructured during the first six years, with penalties and fines restructured in the following four years. If one-third of tax and insurance debt were repaid during the first two years, then 50 percent of penalties and fines would be written off. If the remainder of tax and insurance debt were repaid during the next four years, all penalties and fines would be written off.

The second policy initiative concerns unprofitable farms, which in 2001 still accounted for about one-half of large agricultural enterprises in Russia. Minister of Agriculture Gordeev suggested that the treatment of chronically unprofitable enterprises "requires a different solution" from that in the past, meaning that pumping endless resources into under-performing enterprises could not be continued.[66] In early 2001 a federal plan had taken shape. Agricultural enterprises were grouped into four categories, and each category of farm will define the steps it needs to take to become solvent, ranging from stronger labor discipline, restructuring of debt, and changes in farms' business plans and management—up to and including bankruptcy and liquidation of assets.[67] For chronically unprofitable farms with little chance of repayment of debt, the preferred approach was for neighboring farms that were profitable to acquire the assets of nonprofitable neighbors. The Ministry of Agriculture began working on a plan to distribute the land and property of chronically unprofitable farms. Oversight for the implementation of the plan in general would come from regional committees, composed of creditors. Specific financial plans for each concrete enterprise would be implemented by the group of creditors for that enterprise.[68] In addition, more responsibility for the material and financial support of weak and unprofitable farms would be shifted to regional administrations.

The third policy initiative concerns agricultural credit. The lack of a reliable credit policy and necessary credit institutions contributed to the decline of agricultural output during the 1990s. In mid-2000, a new credit system was announced by Gordeev. The new system provides subsidized interest rates to purchase needed inputs or to fund operations, instead of providing subsidized inputs as in the recent past. The new credit system took effect for the 2001 sowing season.

In this new system, state credits are transferred to Rossel'khozbank (and other banks, including Sberbank and other Moscow banks), which then provide loans directly to food producers at higher interest rates.[69] The federal budget includes a budget line (a preferential credit fund) for the subsidization of seasonal loans. The goal of the system is to increase the effectiveness of credits and provide banks with an incentive to participate, while at the same time providing the seasonal support that food producers need.

In sum, policies designed to address the debt and credit challenges appear to create conditions for financial renewal. Moreover, large agricultural enterprises are likely to see their financial condition improve as a result of protectionist policies from the government, and will benefit from state interventionist policies that provide price supports. Finally, the policy toward unprofitable farms creates a basis for more efficiency and higher levels of production in the future. Thus, the prospects for financial strengthening of large farms and private farms are good, at least in the short term.

Land: Legal and Economic Prospects

Although the Land Code codified the right to buy and sell land, there are several reasons to doubt that this document alone will jump-start the Russian land market and lead to significant societal transformation in the near to mid-term. For example, a large percentage of land remains in municipal ownership, and how much will be sold off remains an open question. Moreover, before a robust land market can arise, a cadastre system has to be developed and implemented, and virtually the whole of Russia needs to be "cadastred." This process could easily take fifteen to twenty years. Finally, it is a long-term proposition to root out and eliminate—if it can be accomplished at all—the corruption, bureaucratism, and petty tyranny of those officials who oversee land transactions.

Once the Land Code was adopted, the Putin government turned its attention to a law regulating agricultural land sales. In February 2002 a draft law on turnover of agricultural land was submitted by the government to the Duma; in March 2002 it was published and began to be debated.[70] During April 2002, as Russia's regions were discussing the law, a sharp debate occurred over whether foreigners would be allowed to buy agricultural land, and if so, how much. In May 2002 the bill passed its first reading in the Duma, and on June 21, 2002, the second reading passed. A few days later, on June 26, 2002, the bill passed its third and final reading by a vote of 258–149.[71] The final version approved by the Duma was sent to the Federation Council, where it passed on July 10, 2002. President Putin signed the bill into law on July 24, 2002. With the stroke of a pen, Putin introduced the first law since 1917 allowing the sale of agricultural land in Russia.

The law consists of four chapters and twenty articles.[72] For all of the extended controversy surrounding agricultural land sales, this law is not a revolutionary document. In fact, one is struck by its conservative nature.[73] It is hard to see how this law will radically transform either rural society or the rural land market, how it will facilitate the transfer of land to the most productive users, or how it will aid in the resolution of the unprofitable farm problem.

Three examples will illustrate the conservative nature of the law: land sales, leased land, and land shares. The sale of agricultural land is governed by Article 8 of the new law. This article regulates the sale of all agricultural land that is not explicitly excluded by the law.[74] According to this article, organs of local government (raion or village administrations) have preferential rights to buy agricultural land. The seller of a land plot is required to submit a written letter, or notification (*izveshcheniye*), to the regional government or, in certain cases, to the raion administration, of his intent to sell his land. This notification should indicate the price of the land and other components of the transaction. The regional government has one month, from the date of receipt of the notification, to exercise its right to purchase the land. If the regional government does not exercise its right to purchase the land, or fails to inform the seller of its intent to acquire the land plot, then the seller has one year to sell the land to a third party at a price not lower than that which was indicated in the letter of intent submitted to the regional government. The one year term starts with the submission of the letter of intent to the regional government.

The selling process is iterative and allows only indirect negotiation over price. If the seller decides to sell the land plot at a price lower than that which was indicated in the original notification, or if terms of the transaction change, the seller is obligated to submit a new *izveshcheniye*, and again the local administration has a month to exercise its right of first refusal. To illustrate, say a person wanted to sell his land plot for price "A." The local government has one month to exercise its right of first refusal at the stated price. If it does not want the land at that price, then the seller can try to sell it to someone else for up to a year, but not at a lower price than indicated in the notification. If the land does not sell, the seller can lower the price to price "B," and again the local government has one month to exercise its right of first refusal at the new, lower price. It is not stated in the law, but presumably the seller does not have to wait a year before withdrawing the land plot from the open "market," but can change the price and offer it to the local government at any point prior to the expiration of a year.[75] Land that is not purchased by a local government body may be sold to third persons one on one, or through auction or competitive bidding.

Article 9 regulates the leasing of agricultural land. Point 3 of this article states that the amount of agricultural land that may be leased is not limited in size; however, the term of the lease may not exceed forty-nine years. Point 2 of the article states that land in the form of land shares may be leased, either from all the owners of the shares, or from one of them who has power of attorney to act on the others' behalf. Article 10 allows municipal land that has been leased to be converted to private ownership upon the expiration of the lease term, according to the process described above in Article 8. Article 10, point 4 allows the purchase of municipal land after a three-year period dating from the conclusion of the lease agreement. In this case, the lessee submits an application to the regional government or local administration, which makes a decision within a two-week period whether or not to grant the request. If the decision is positive, the leased land may be purchased at the current "market value."

Finally, Article 12 of the law concerns the sale of land shares held by farm members. The owner of a land share is obligated to inform other owners (members of the farm) of his intent to sell his land shares. The notification must be in the form of a letter, or published in a source of mass information, as defined by the regional authorities. The rest of the farm members have one month to exercise their right of first refusal. If the other members turn down the opportunity to purchase the land shares, then the owner who wishes to sell has to notify the regional or local administration of his intent to sell, to define the price, terms, conditions, etc. In short, the same iterative process described above in Article 8 comes into play. If the local government declines to exercise its right of first refusal within a month, the owner of the land shares has a year to sell to a third party, but again at a price not lower than that which was indicated in the notification. Essentially, a seller of land shares has to first offer it to other members of the farm, then to the local government, before he can sell to a third party. Thus, these three examples show that government organs will be heavily involved in the land market, but not in a good way. The selling process is cumbersome, the process is ripe for corruption and bribery, and the influence of supply and demand is diminished.

Socioeconomic and Demographic Prospects

We have chosen to discuss socioeconomic and demographic prospects last because these policy challenges are likely the most daunting. Whereas the other rural policy challenges either are in the beginning stages of revival or have prospects for revival, our assessment of socioeconomic and demographic factors is not as optimistic. First, there is a need to increase rural incomes and decrease rural poverty, which in turn will have other positive effects such as raising rural standards of living, building sustainable rural communities, and providing incentives to remain employed in the agricultural sector. However, the problem is how to increase rural incomes. Rural incomes are linked to food prices. There is little political will to see food prices for urban consumers rise substantially. Moreover, if food production continues to grow, either exports must increase or downward pressure will be exerted on domestic food prices, thereby lowering the incomes of farms and their personnel. However, Russia lacks adequate export facilities, and the protectionist policies of its neighbors call into doubt how much food exports can be increased. Agricultural Minister Gordeev has made clear the government's intention to "regulate" the food market—specifically in grain— by intervening and buying up surpluses; in effect providing price support through government purchases.[76] While this action benefits grain growers and props up their incomes, the question arises, however, how interventionist the Putin government is willing to become—how many commodities will benefit from similar interventions? In 2001, a special intervention fund was created and funded with two billion rubles (about 69 million dollars at 2001 exchange rates). However, in 2001 agricultural production was valued at about 1 trillion rubles, and thus choices will have to be made as to which commodities will benefit.

The second main problem is rebuilding rural infrastructure. The fulfillment of this requirement depends upon the overall condition of the economy, which in turn is linked to the price of Russia's primary export commodities—oil and gas. In short, even assuming the political will to allocate more federal funds to rural capital investment, the economic prospects are unsure at best.

The third main problem is how to rebuild the base of human capital. The prospects are not good. Rural Russia is aging, and the migratory influx experienced in the early 1990s has abated. Also during the early 1990s, the private farm movement attracted many urbanites into agricultural production. Today, however, the prospect of becoming a productive (and affluent) private farmer—an idea that was so attractive during the first half of the 1990s—has faded in appeal. The private farming experiment has not been very successful in Russia, and as a consequence the Putin government has turned its attention back to large farms. Moreover, rural schools are substandard, as is rural infrastructure in general. Combining high rates of rural unemployment, limited professional opportunities, low incomes and standards of living, as well as limited social outlets and personal opportunities for marriage, it is hard to imagine how the unattractiveness of rural life can be overcome in the near to mid-term. In short, the socioeconomic and demographic challenges are likely to continue for the foreseeable future.

Notes

1. See Stephen K. Wegren, "Risk Environments and the Future of Russian Private Farming," *Current Politics and Economics of Russia, Eastern and Central Europe*, vol. 16, no. 2 (2001), p. 130.

2. *Rossiia v tsifrakh* (Moscow: Goskomstat, 2001), p. 199. The 43 percent actually represents an improvement from the nadir of 39 percent in 1998. After the financial collapse of 1998, domestic producers' food products became more price competitive with imports, and demand for domestic products increased.

3. For data on regional production variations, see *Sel'skokhoziaistvennaia deiatel'nost' khoziaistv naseleniia v Rossii* (Moscow: Goskomstat, 1999), pp. 72–152.

4. *Rossiia v tsifrakh*, p. 200.

5. See Stephen K. Wegren, "The Russian Food Problem: Domestic and Foreign Consequences," *Problems of Post-Communism*, vol. 47, no. 1 (January-February 2000), p. 40; and *Prodovol'stvennyi rynok Rossii* (Moscow: Goskomstat, 2000), p. 56.

6. *Sel'skoe khoziaistvo v Rossii* (Moscow: Goskomstat, 2000), p. 135.

7. See *Potreblenie produktov pitaniia v domashnikh khoziaistvakh v 1999–2000 gg (po itogam vyborochnogo obsledovaniia biudzhetov domashnikh khoziaistv)* (Moscow: Goskomstat, 2001), pp. 12–14.

8. See Stephen K. Wegren, "Russian Agrarian Policy Under Putin," *Post-Soviet Geography and Economics*, vol. 43, no. 1 (January–February 2002), pp. 29–32.

9. Grain imports were a fraction of the levels during the latter Soviet period. For example, during 1981–1985 the Soviet Union imported an average of more than 40 million tons of grain a year; during 1986–90, the Soviets imported an average of more than 35 million tons annually, or roughly 15 percent of total grain availability. In the 1990s, grain imported averaged around 3–4 million tons.

10. Percentages of supply were calculated from *Rossiiskii statisticheskii ezhegodnik* (Moscow: Goskomstat, 2001), pp. 416, 611, 614.

11. *Rossiia v tsifrakh*, pp. 366–67.

12. *Krest'ianskaia rossiia*, no. 13 (2001), p. 1.

13. There is significant bureaucratic support for protection of domestic producers beyond the Ministry of Agriculture. Advocates for food security draw support from large farms and private farmers, from food-producing regions, from food processors, and from various agricultural interest groups such as the Meat Union, the Grain Union, the Sugar Union, as well as the Agrarian Party of Russia.

14. A. V. Gordeev, A. I. Altukhov, and D. F. Vermel', "Prodovol'stvennaia bezopasnost' Rossii: sostoianie i meri obespecheniia," *Ekonomika sel'skokhoziiaistvennykh i pererabatyvaiushchikh predpriiatii*, no. 10 (October 1998), p. 10.

15. Available at http://www.kommersant.ru/Docs/high-priority-task.htm.

16. *RFE/RL Newsline*, vol. 6, no. 71, part 1, April 16, 2002, pp. 1–2; ibid., vol. 6, no. 93, part 1, May 20, 2002.

17. *Sel'skaia zhizn'*, March 12, 2002, p. 1.

18. *The Russian Journal*, March 18, 2002, p. 3.

19. *The Financial Times*, March 31–April 1, 2001, p. 3.

20. *Agropromyshlennyi kompleks Rossii* (Moscow: Goskomstat, 2001), p. 55.

21. *Krest'ianskie vedomosti*, nos. 15–16 (2001), p. 2.

22. A. I. Manellia, "Sel'skoe khoziaistvo Rossii v 2001 godu," *Ekonomika sel'skokhoziaistvennykh i pererabatyvaiushchikh predpriiati*, no. 2 (February 2002), p. 36.

23. *Agropromyshlennyi kompleks Rossii*, p. 57.

24. Rural capital investments increasing slightly to 3 percent in 1998 and 2.9 percent in 1999. In 2000, agriculture received 2.6 percent of capital investments. *Rossiiskii statisticheskii ezhegodnik* (2001), p. 570.

25. L. Bondarenko, "Sostoianie sotsial'no-trudovoi sfery sela," *Voprosy ekonomiki*, no. 7 (July 2000), p. 72.

26. *Sotsial'no-ekonomicheskoe polozhenie Rossii*, no. 12 (December 1995), p. 65.

27. "Sel'skoe khoziaistvo Rossiiskoi Federatsii v 1996–2000 godakh," *APK: ekonomika, upravlenie*, no. 11 (November 2001), p. 17.

28. See Alfred B. Evans, Jr., "Rural Russia: Survival in a Demodernizing Economy," paper presented at the Western Slavic Association and Western Social Science Association, Albuquerque, New Mexico, April 10–13, 2002.

29. Stephen K. Wegren, "Rural Politics and Agrarian Reform in Russia," *Problems of Post-Communism*, vol. 43, no. 1 (January–February 1996), p. 26.

30. A. V. Petrikov, "Sotsial'nye problemy Rossiiskoi derevni," *Ekonomika sel'skokhoziaistvennykh i pererabatyvaiushchikh predpriiatii*, no. 3 (March 1999), p. 38.

31. *Rossiiskii statisticheskii ezhegodnik* (Moscow: Goskomstat, 2000), p. 172.

32. Data are cited in Evans, "Rural Russia: Survival in a Demodernizing Economy."

33. Petrikov, "Sotsial'nye problemy Rossiiskoi derevni," p. 38

34. *Narodnoe khoziaistvo Rossiiskoi Federatsii. 1992* (Moscow: Goskomstat, 1992), pp. 405–408.

35. *Trud i zaniatost' v Rossii* (Moscow: Goskomstat, 1995), p. 49. This was a trend that continued until the end of the decade, as in 1998, a worker on a large agricultural enterprise earned 39 percent the average monthly income of an industrial worker, and in 1999, 34 percent. *Trud i zaniatost' v Rossii* (Moscow: Goskomstat, 1999), p. 309; and *Rossiiskii statisticheskii ezhegodnik* (2000), p. 155.

36. L. V. Bondarenko, "Sostoianie sotsial'no-trudovoi sfery sela," p. 69.

37. Official Russian data for the nation depict falling rural standards of living. Some Russian commentators suggest that due to family land plots, rural standards of living declined less than those of urban families. A team of American and Russian sociologists found, based on a panel study of three villlages during a three-year time period, some evidence that the material well-being of rural families increased. See David O'Brien, Valeri V. Patskiorkovski, and Larry Dershem, *Household Capital and the Agrarian Problem in Russia* (Aldershot, UK: Ashgate, 2000).

38. *Demograficheskii ezhegodnik Rossii* (Moscow: Goskomstat, 2001), p. 83.

39. For a review, see Stephen K. Wegren and Vladimir R. Belen'kiy, "Change in Land Relations: The Russian Land Market," in David J. O'Brien and Stephen K. Wegren, eds., *Rural Reform in Post-Soviet Russia* (Washington, DC: Woodrow Wilson Center/Johns Hopkins University Press, 2002), chapter 4.

40. See Zvi Lerman and Karen Brooks, "Russia's Legal Framework for Land Reform and Farm Restructuring," *Problems of Post-Communism*, vol. 43, no. 6 (November–December 1996), pp. 48–58.

41. *Krest'ianskie vedomosti*, nos. 3–4 (2001), p. 2.

42. *RFE/RL Political Weekly*, vol. 1, no. 10 (March 26, 2001).

43. *Sel'skaia zhizn'*, July 17, 2001, p. 1.

44. *Krest'ianskaia rossiia*, no. 40 (2001), p. 2.

45. "O vvedenii v deistvie Zemel'nogo kodeksa Rossiiskoi Federatsii," *Sobranie zakonodatel'stva Rossiiskoi Federatsii*, no. 44 (October 29, 2001), pp. 9236–9241.

46. See "Zemel'nyi Kodeks Rossiiskoi Federatsii," *Sobranie zakonodatel'stva Rossiiskoi Federatsii*, no. 44 (October 29, 2001), pp. 9175–9236.

47. These points are drawn from Vladimir R. Belen'kiy, "Socio-Economic Consequences of the Land Code Entering into Force," paper presented at "Russian Security Challenges in the 21st Century," Southern Methodist University, April 6, 2002.

48. It is alleged that relevant experts were not consulted for their advice, which is one reason the Land Code was criticized.

49. *Gosudarstvennyi (natsional'nyi) doklad o sostoianii i ispol'zovanii zemel' Rossiiskoi Federatsii v 2000 godu* (Moscow: Federal Land Cadastre Service of Russia, 2001), p. 121.

50. Ibid.

51. *Rossiia v tsrifrakh*, p. 106.

52. Wegren and Belen'kiy, "Change in Land Relations: The Russian Land Market," in O'Brien and Wegren, eds., *Rural Reform in Post-Soviet Russia*, chapter 4, Table 5.

53. *Krest'ianskie vedomosti*, no. 7, 2000, p. 2.

54. Text of the July 27, 2000, speech can be downloaded from the website of the Russian Ministry of Agriculture: http://www.aris.ru/MSHP/vistgord.html.

55. *Agropromyshlennyi kompleks Rossii*, p. 14; and "Sel'skoe khoziaistvo Rossii v 2001 godu (ekonomicheskii obzor)," *APK: ekonomika, upravlenie*, no. 3 (March 2002), p. 20.

56. "Sel'skoe khoziaistvo Rossiiskoi Federatsii v 1996–2000 godakh (ekonomicheskii obzor)," p. 16; and "Sel'skoe khoziaistvo Rossii v 2001 godu (ekonomicheskii obzor)," p. 20.

57. C. Peter Timmer, Walter P. Falcon, and Scott R. Pearson, *Food Policy Analysis* (Baltimore: The World Bank/Johns Hopkins University Press, 1983), p. 56.

58. Food consumption data may be found in *Potreblenie produktov pitaniia v domanshnikh khoziaistvakh v 1997–1999 gg (po itogam vyborochnogo obsledovaniia biudzhetov domashnikh khoziaistv)* (Moscow: Goskomstat, 2001), pp. 12–14; and *Potreblenie produktov pitaniia v domanshnikh khoziaistvakh v 1999–2000 gg*, pp. 12–14.

59. V. Semenov, V. "Novyi kurs agrarnoi politiki," *Ekonomist*, no. 1 (January 1999), p. 14.

60. *Krest'ianskie vedomosti*, no. 3 (2000), p. 2.

61. *Krest'ianskie vedomosti*, nos. 19–20 (2001), p. 2.

62. See K. G. Borodin, "Netarifnye bar'ery v torgovle agroprodovol'stvennoi produktsiei mezhdu stranami SNG," *Ekonomika sel'skokhoziaistvennykh i pererabatyvaiushchikh predpriiatii*, no. 3 (March 2002), pp. 14–16; and K. Borodin, "Ustranit' bar'ery agrarnogo rynka SNG," *APK: ekonomika, upravlenie*, no. 3 (March 2002), pp. 42–48.

63. This was confirmed in a round table interview with Gordeev, published in early 2002. See "Glavnye napravleniia stabilizatsii i razvitiia APK," *Ekonomika sel'skokhoziaistvennykh i pererabatyvaiushchikh predpriiatii*, no. 3 (March 2002), p. 8.

64. This section draws from Wegren, "Russian Agrarian Policy Under Putin," pp. 32–38.

65. "O poriadke i usloviiakh provedeniia v 2001 godu restruturizatsii prosrochennoi zadolzhennosti (osnovnogo dolga i protsentov, peney i shtrafov) sel'skokhoziaistvennykh predpriiatii i organizatsii," *Sobranie zakonodatel'stva Rossiiskoi Federatsii*, no. 25 (June 18, 2001), pp. 5018–5022.

66. *Sel'skaia zhizn'*, January 30, 2001, p. 2.

67. The four categories were: (1) profitable farms without debt (22 percent of all farms); (2) profitable farms based on current production (17 percent of all farms); (3) unprofitable farms based on current production but which have the potential to repay their debts (34 percent of all farms); and (4) unprofitable farms with no potential to repay their debts (27 percent of all farms).

68. *Sel'skaia zhizn'*, January 30, 2001, p. 2.

69. For more detail on how the system works, see *Krest'ianskie vedomosti*, nos. 13–14 (2001), p. 4.

70. *Sel'skaia zhizn'*, March 19, 2002, pp. 1–2.

71. *RFE/RL Newsline*, vol. 6, no. 119, part 1, June 26, 2002.

72. The full text of the law is found in *Sel'skaya zhizn'*, August 1–7, 2002, pp. 8–9.

73. For an analysis of the law, see Stephen K. Wegren, "Observations on Russia's New Agricultural Land Legislation," *Eurasian Geography and Economics*, vol. 43, no, 8 (2002).

74. The sales process described in the article does *not* apply to agricultural land owned by citizens who use it for individual housing, construction of garages, conducting subsidiary and dacha agricultural production, collective gardens and orchards,

and also assorted buildings. It should be noted that these types of land plots—which tend to be very small—comprise the bulk of purchase transactions among individuals, so this submarket remains outside the scope of the new law.

75. The law also does not state explicitly what would happen if a local government did not want the land and the owner could not sell it to a third party, at least at his asking price. Article 6, point 3, indicates that land must be used properly or it can be confiscated, so presumably this implies that the owner must continue to cultivate his land or risk losing it. Remember also that the Land Code stipulates that land not used for three years risks confiscation.

76. "Glavnye napravleniia stabilizatsii i razvitiia APK," p. 8.

For Further Reading

Stephen K. Wegren, *Agriculture and the State in Soviet and Post-Soviet Russia* (Pittsburgh: University of Pittsburgh Press, 1998).

Stephen K. Wegren, "The Russian Food Problem: Domestic and Foreign Consequences," *Problems of Post-Communism*, vol. 47, no. 1 (January–February 2000).

David O'Brien, Valeri V. Patskiorkovski, and Larry Dershem, *Household Capital and the Agrarian Problem in Russia* (Aldershot, UK: Ashgate, 2000).

Stephen K. Wegren, "Risk Environments and the Future of Russian Private Farming," *Current Politics and Economics of Russia, Eastern and Central Europe*, vol. 16, no. 2 (2001).

Stephen K. Wegren, "Russian Agrarian Policy Under Putin," *Post-Soviet Geography and Economics*, vol. 43, no. 1 (January–February 2002).

David J. O'Brien and Stephen K. Wegren, eds., *Rural Reform in Post-Soviet Russia* (Washington, DC: Woodrow Wilson Center/Johns Hopkins University Press, 2002).

13

Environmental Policy Challenges

Craig ZumBrunnen and
Nathaniel Trumbull

Russia's human population has interacted with Russia's environmental diversity and vastness to create several ironies, many of them tragic. On the one hand, Russia still contains some of the world's largest expanses that have barely been impacted by humans. On the other hand, Russia contains some of the most environmentally degraded and polluted air, water, and landscapes anywhere on the planet. Accordingly, Russia's environmental policy challenges for the twenty-first century are legend and physically, spatially, and institutionally exacerbated by the legacy of decades of Soviet policies, priorities, and practices that placed low value on the rational use of the environment and its resources, including the protection and preservation of ecological health and diversity. In this chapter we present and discuss many of the major geographical dimensions and ideological, political, economic, and various institutional factors that have generated the myriad of Russia's current and future environmental policy challenges. The evidence strongly indicates that the current situation and trends are not sanguine for the health of either Russia's physical environment or her human population.

Survey of the Physical Nature of Russia's Environmental Problems

What exactly has been and is the physical nature of Russia's environmental suffering? Space does not allow for even a cursory overview of all the major environmental problems currently facing the Russian Federation's political leadership, policy makers, and citizens. We are thus left here with room only to list by name the major types of physical and ecological environmental

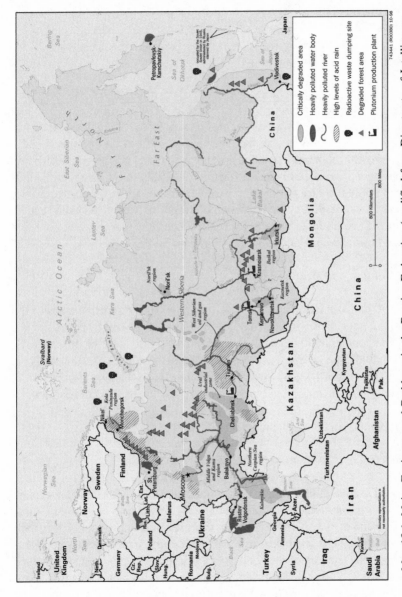

Map 13.1. Major Environmental Problems in the Russian Federation (modified from Directorate of Intelligence, *Handbook of International Economic Statistics, 1998*, Washington, DC: US Government Printing Office, 1998).

problems, skipping over their main causes, trends, impacts, severity, and spatial patterns. However, due to the rather well-established scientific link between levels of air pollution and the incidence of human health problems, slightly more detail about Russian air pollution problems will be included (see Map 13.1).

Water Quality and Water Availability

Very few of the medium-size to large-size rivers and lakes of Russia remain free from pollution, even in the Far North, Siberia, and the Far East. The most severely polluted waterways are located in the Volga-Caspian Basin and the Ob'-Irtysh Basin, and the latter basin's waterways suffer especially from petroleum contamination. Freshwater availability is problematic in many regions, and a significantly larger fraction of Russia's best agricultural lands are susceptible to drought than was the relative case for the former Soviet Union as a whole. A very large fraction of the entire Russian coastline and its estuaries have polluted waters, especially along the Barents Sea coastline and the coastal waters of the Russian Far East.[1]

Soil, Habitat, and Wildlife

Both soil contamination and soil erosion are serious. Extensive areas of the country are negatively impacted by the pollution of the soil by heavy metals. Pesticide contamination extends throughout the entire range of Russian agricultural lands, but is especially severe in the North Caucasus, Central Chernozem (Black Earth), and lower Volga River basin. Habitat destruction is especially significant in the logging areas of northwest Russia, southern Siberia, and the Far East. Lucrative poaching has increased the number and geographical extend of endangered species.[2]

Radioactive Contamination and Hazards

Radioactive contamination and the risks of nuclear submarines and power plant accidents are very real. For example, there are reported to currently be some two thousand organizations in Moscow that used radioactive materials in their work, including eleven nuclear reactors. The largest nuclear-wastes deposit is located on the territory of the Kurchatovsky Institute nuclear research facility, and the Ministry for Nuclear Power acknowledges this fact. There was no radiation control system at all in Moscow until 1960. As a result, dangerously radioactive wastes were buried throughout the city. Some 70 percent of the radioactively contaminated sites are often found during the excavations for construction projects. Over the past twenty-five years the Moscow City government has liquidated over 1,350 such contaminated sites and has removed over 930 tons of radioactive wastes.[3] And in terms of severity, Moscow is far less radioactively contaminated than other regions in the vicinity of Krasnoyarsk, Cheliabinsk, Lake Karachai, Miass in the Urals, and elsewhere. A description and map of the radioactive contamination "hot spots" in Russia paints a truly chilling portrait for current and future Russian citizens.[4]

Natural Hazards

Russian environmental policies also will need to address numerous natural hazards such as earthquakes, volcanic eruptions, tsunamis, droughts, floods, and forest fires. The link between human health problems and environmental pollution seems tragically well established in Russia.[5] Even simply as a list of environmental problems crying out for sound policy responses, this listing is very far from complete, and readers are referred to nearly all of the references cited at the end of this chapter for more specific details. However, before proceeding with a historical retrospective of Soviet/Russian environmental policy, we have decided to include at least a bit more detail about one of Russia's more serious environmental problems, air pollution.

Air Pollution Levels

Air quality has long been and remains a major problem throughout the urbanized realms. One of the major changes in the post-Soviet era is that the air quality in the cities of European Russia that had experienced a quantum improvement in the late seventies and eighties when natural gas replaced coal and oil for many industrial and domestic uses (especially fueling boilers for space heating) has again deteriorated due to exhaust gases of the automobiles whose number has been swelled by millions of used and poorly tuned imports from elsewhere in Europe and Japan. During the Soviet era, gaseous and aerosol releases from the thousands of industrial smokestacks dotting the landscape of the "fertile triangle" (the territory enclosed by an imaginary triangle with vertices at then Leningrad, Irkutsk, and Odessa in Ukraine) belched millions of tons of contaminants into the atmosphere annually, with one of the worst places and most deleteriously ecologically impacted regions being the metallurgical center of Noril'sk in northern East Siberia.[6] At Noril'sk, stationary sources annually emitted more tonnage of air pollutants than the combined tonnage from the smokestacks of the four next worse urban industrial complexes, Novokuznetsk, Magnitorgorsk, Cherpovets, and

Figure 13.1. **Discharge of Polluting Substances into the Atmosphere
of the Russian Federation by Branch of Industry in 1999** (in %)

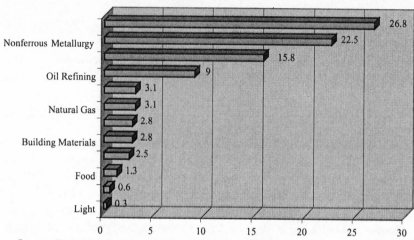

Source: S. A. Ushakova and Ia. G. Katsa, eds., *Ekologicheskoe sostoianie territorii
Rossii* (Moscow: Academia, 2001), p. 16.

Lipetsk, all major metallurgical centers.[7] In 1999 approximately 18,600 Russian enterprises had a total of 955,000 stationary sources from which 79.6
million metric tons of air pollutants were emitted. Of this mass, 15.7 million
metric tons were discharged without any purification, but 63.9 million metric tons were treated at purification installations where 61.1 million metric
tons were intercepted and neutralized. During the 1990–1999, period the
discharge of several of the major pollutants decreased from 5 to 49 percent,
but this was due overwhelmingly to the general decline of industrial output.[8]
Figure 13.1 portrays the share of total industrial air pollution released by
industrial branch.

Rosgidromet (Russian Hydrometeorological Service) regularly monitors
219 cities and populated places in the Russian Federation from 621 observation posts. The concentrations of from 5 to 25 substances are routinely measured in the largest cities. In 1999 Russia had 195 cities with a combined
population of 64.5 million (44 percent of Russia's population) that had average annual concentrations of one to several atmospheric pollutants that exceeded their respective *PDK*s (Maximum Permissible Concentrations). A total
of 79 Russian cities are listed as having overall high air pollution levels. The
industrially generated air pollution in the cities consists mainly of high concentrations of benzpyrene (a known carcinogen), suspended particulates, NO_2,
hydrocarbons, and formaldehyde.[9] Table 13.1 lists the 22 Russian cities currently experiencing the highest levels of air pollution.

Table 13.1

Cities with the Highest Levels of Polluted Air in 1999

City	Substances determining the high level of polluted air
Balakovo	hydrocarbons, formaldehyde, NO_2
Biisk	formaldehyde, suspended particulates, NO_2
Bratsk	NO_2, formaldehyde, hydrogen fluoride, hydrocarbons
Ekaterinburg	formaldehyde, benzpyrene, acrolein
Irkutsk	formaldehyde, suspended particulates, NO_2
Kemerovo	hydrocarbons, ammayak, formaldehyde, soot
Krasnoyarsk	benzpyrene, suspended particulates, chlorine
Krasnodar	phenol, formaldehyde, suspended particulates
Lipetsk	phenol, ammonia, formaldehyde, NO_2
Magadan	phenol, formaldehyde, NO_2
Magnitogorsk	benzpyrene, phenol, suspended particulates
Moscow (region)	ammonia, NO_2, formaldehyde
Novokuznetsk	formaldehyde, suspended particulates, hydrogen fluoride, NO_2
Novorossiisk	NO_2, benzpyrene, suspended particulates
Omsk	formaldehyde, acetaldehyde, soot
Rostov-na-Donu	NO_2, formaldehyde, suspended particulates
Selenginsk	formaldehyde, phenol, hydrocarbons, methylmercaptan
Tiumen	suspended particulates, formaldehyde, lead
Ulan-Ude	suspended particulates, formaldehyde, NO_2
Khabarovsk	benzpyrene, formaldehyde, SO_2, NO_2, formaldehyde, ammonia
Chita	benzpyrene, formaldehyde, suspended particulates, NO_2
Iuzhno-Sakhalinsk	soot, suspended particulates, NO_2

Source: S. A. Ushakova and Ia. G. Katsa, eds., *Ekologicheskoe sostoianie territorii Rossii* (Moscow: Academia, 2001), pp. 13–14.

As noted previously, the ecological problems in Russia's urbanized regions have become exacerbated as a result of the rather dramatic increased "automobilization" of cities since the breakup of the Soviet Union. In some individual cases, automobile ownership has reached 300 cars per 1,000 inhabitants. Such situations exist in the Moscow, Saint Petersburg, Saratov, Cheliabinsk, and Novosibirsk urban agglomerations. Air pollution from automobile exhaust gases continues to grow and currently accounts for 50 percent or more of the total mass of atmospheric discharges in Russia's largest cites. In Moscow, auto exhausts contribute 93 percent of the total contaminated airborne discharges.[10] This necessarily cursory description and discussion of Russian air pollution by no means is even intended to provide a "clear" picture of complexities of air quality problems in Russia. Nonetheless, we need to turn to a historical retrospective on environmental protection problems in Russia.

The Historical Legacy

Soviet Legacy and Inheritance

Academic discussions of environmental issues had already emerged in Soviet scientific circles in the1960s.[11] During the late Soviet period, a number of precursors to Russia's state environmental protection in the 1990s had emerged. An elaborate set of environmental laws and several agencies required a detailed reporting of environmental conditions and violations.[12] But those laws existed largely on paper rather than in practice. Legislation such as the 1960 Law on Air Protection and the 1972 Water Code established seemingly strict norms, in some cases stricter than in the West, but in practice those norms were not widely enforced.[13] Soviet factory managers often misreported water effluent and air pollution data. A number of Western specialists have demonstrated that even officially published environmental statistics could be largely inaccurate.[14] The late Soviet period was also a time of relative détente and unprecedented numbers of exchanges between Soviet and Western scientists. Yet knowledge about environmental issues was considered specialized knowledge and not for public consumption or analysis. To an unfortunate degree, the Soviet public at large remained without access to scientific knowledge of environmental issues, their health effects, and the extent of environmental degradation in their own country.[15] Soviet officials justified their silence on the topic under the pretense of not wishing to alarm the public with information that the public was seen as unprepared to interpret scientifically.

A number of highly visible cases of environmental disruption nevertheless came to be discussed in the press in the second half of the 1980s as press censorship was lifted. The old environmental cause célèbre from the 1960s, Lake Baikal, once again took center stage.[16] Even the Soviet planners' greatest ambitions for industrialization of the region could not entirely outweigh the threat to the lake's large number of unique biological species. The total number of planned paper plant projects to be built on the lake, especially on the northern shore where the Baikal-Amur-Magistral (BAM) railway line passed, was eventually revised and lowered, and a ban was imposed on the movement of logs on the lake itself.[17] A similar ban on log floating on Baikal's tributaries had been imposed by decree in 1960, but obviously was not effective or enforced.[18] Debate over construction of a flood barrier project in Leningrad also captured national attention and debate. The proposed reversal of Siberian rivers in order to make them flow south to the arid lands of Central Asia, which had a long history of debate, much of it translated by Ted Shabad for *Soviet Geography Review and Translation*, once again became part of public discussion.[19] The 1986 Chernobyl incident tragically reinforced the conclusion among the Soviet public that the Soviet government had not placed public health first. Indeed, it has been argued that environmental degradation and its accompanying health consequences played an important role in discrediting the legitimacy of the Soviet state in the eyes of its own citizens.[20]

Placing the Blame on Capitalism's Institutions

Prior to Gorbachev, a common Soviet refrain was that environmental problems were the natural outcomes of capitalism's institutional triad of private property, the profit motive, and "free market" competition, all of which create powerful incentives for individuals and firms to generate environmental externalities or social costs by discarding their unwanted industrial, mining, forestry, and agricultural by-products into the air, water, and land of the surrounding environment, thereby lowering their production costs and increasing their entrepreneurial competitiveness in the marketplace. Indeed, Western economic theorists since the time of Pigou's writings in the nineteenth century developed and honed such arguments. Soviet leaders pointed to the work of Pigou to affirm this perspective to Western observers and their own citizens as environmental problems began to surface in the former Soviet Union. They attributed these problems to vestiges of capitalism rather than to shortcomings of the Soviet command economy or rapid economic growth in general. Accordingly, while the environmental movement in the West was burgeoning rapidly even before the first Earth Day in 1970, Soviet and Marxist theoreticians and policy makers were arguing that Western environmental problems provided convincing empirical evidence that capitalism and its profit motive breed environmental disruption and destruction.

The Soviet Theoretical Ability to Prevent Environmental Problems

The Soviet Union's leadership's counterargument was that lack of capitalism with its private property rights, private profit motive, and free markets protected it from serious environmental problems. For example, some four decades ago, the Soviet academician, Professor N. A. Gladkov, asserted that:

> In the Soviet Union it is not as if there were need of special measures for the protection of nature as the very structure of Soviet society founded on

a planned economy and on the absence of private property ensures the preservation and growth of the natural world.[21]

The Soviets thus attempted to build implicitly, if not explicitly, their own environmental policy from their mainly improvised institutional "troika." The "troika" comprised: (1) state ownership and collective ownership of property based on the labor theory of value replacing private property, (2) a system of administrative and employee bonuses to replace the profit motive, and (3) central planning and allocation of resources and products to replace the market. Igor Petryanov, a former editor of *Khimiia i zhizn'* [Chemistry and Life] and a Soviet scientist, argued, "that the prerequisites for the most prudent use of resources of the biosphere are in the Soviet social system itself" and that a technologically "planned" society rather than a biologically "open democratic society" was the best hope for human ecological survival.[22] A 1979 article in *Ekonomicheskaia gazeta* was more blunt in its claim that:

> History shows us that the sole objective of capitalist society is gain—the maximum possible profit. This attitude predetermines a destructive impact on nature. It's not only Marxists who point to capitalism as responsible for the destruction of the biosphere. All objective scholars realize that, despite certain rays of hope, the ecological situation remains alarming, and there is only one reason why these problems persist—the search for a quick profit. ... In contrast to capitalist production, which plunders nature, the socialist system, based on a planned economy, ensures an improvement of the environment, providing genuine guarantees that mankind will be able to ward off the threat to the ecology.[23]

On the contrary, however, for over thirty years, from 1970 to the present, Western scholars, as well as many Soviet/Russian scholars, scientists and policy makers, have pointed to a multitude of serious and pervasive Soviet environmental problems that were not prevented, and, in fact, were engendered or exacerbated by the Soviet command economy's institutional "troika."

The Soviet Practical Propensity to Pollution and Degrade the Environment

The Soviet "troika" with its three Soviet institutional "horses" emphatically did not create a stable and wise environmental policy sleigh upon which to controllably slide forward toward wise stewardship of the environment. In the labor theory of value, resources from the environment are considered to be free goods, and their value represents only embedded labor. The application of labor theory, did, indeed, generate absurdly low planned or "assigned" prices on environmental resources that led to enormous waste and inefficiencies in the Soviet economy. The managerial reward system with its reliance on bonuses for plan fulfillment transmogrified capitalist profit maximization into equally environmentally damaging incentives for production maximization. The best that can be said for the belief that central planning and allocation of resources would eliminate and preclude all environmental problems is that it was profoundly naïve. Other command economy impediments and shortcomings included: (1) its vertical—as opposed to horizontal—information flow structure with its incentives to distort and hide critical information needed for rational "planned" decision making, (2) a myriad of administrative failures, (3) nonaligned social preferences amongst the party *apparatchiki*, *nomenklatura*, scientists, and average citizens, (4) the failure of political pressure by independent conservation/environmentalist forces, (5) an ideology —partially fueled by xenophobic fears—bent on rapid economic growth at any cost, and, more difficult to prove, (6) the sheer geographical vastness of the Soviet Union, which led to an insidious form of complacency regarding the environment.[24]

Even more naïve would be the opposite conclusion that the lack of planning in some nontransparent manner translates into sound environmental policies! Obviously, the manner in which humans interact with the environment as they individually and collectively pursue their economic lives is critically important, and both crosscuts and must inform all institutional frameworks. Different institutions and policy instruments can yield similar harmful environmental results as environmental destruction case studies from both capitalist and command economies abundantly and tragically reveal. Thus, as the nation transitions away from a command economy, the fundamental Russian environmental policy challenge from an institutional capacity building perspective is to adopt policies and policy tools that stem environmental deterioration and foster ecologic-economic sustainability. Concomitantly, the new Russian Federation and its citizens must invest in and create the human, technological, and financial will to implement new scientifically, socially, and economically sound, environmentally related policies and actions. Unfortunately, the Russian Federation's economy seems to be moving toward several unbridled "Wild West" forms of economic institutions and behaviors that will only hark back to a Pigouian analysis, or resurgent capitalism, as the institutional explanation of environmental problems while the environment continues to suffer, threatening both ecological and human health.

USSR to Russian Federation: Environmental Policy in the Transition Years

Legal Framework and Policy Tools of State Environmental Protection in Russia

The development of state environmental protection in Russia since the collapse of the Soviet Union has closely paralleled the process of emergence, growth, and consolidation of Russian democratization. The period has been characterized by increasing attention to state environmental protection during the *glasnost* period in the mid- and late 1980s, the strengthening of that protection during the 1990s, a devolution toward regional environmental protection responsibility in the mid-1990s, and the eventual consolidation of state environmental protection into Putin's "strengthening of vertical power" (*ukreplenie vlastnoi vertikali*). State environmental protection during this period has been marked by remarkable evolution and change, but also by continuity.[25] After more than a decade of reform in Russia, state environmental protection resembles more the Soviet state's approach to exploitation of the natural environment than it has in any period since the collapse of the Soviet Union. The periodization that follows traces the institutional context of the developments (and the subsequent near dismantling) of state environmental protection in Russia from the *glasnost* period to the present time.

Emerging Focus on the Environment

Public response to environmental degradation on the whole remained stifled until the last years of the Soviet Union as a result of the paucity of available information on the subject. As one environmental specialist has written: "The seventy-three-year history is a history of systematic misinformation on the environmental situation in Russia."[26] Criticism of the Soviet government's lack of divulgence of public information turned out to be well founded when the floodgates of information on the actual state of the environment in Russia were opened. Heated debate and criticism in the Congresses of Peoples' Deputies in 1989 and 1999 came to focus to a large degree on environmental degradation and especially its health consequences. The Soviet regime's cavalier approach in its exploitation of the natural environment began to be fully revealed to the public. Public opinion polls of this period showed that the environment ranked second or third among the problems that most concerned the nation's citizens.[27]

As a result of the growing attention on the degradation of the Soviet Union's natural environment, a January 1988 Soviet government decree established the USSR Committee on Environmental Protection and Natural Resource Use (*Goskompriroda*). It was to replace a much weaker USSR Council of Ministers' Commission on Environmental Protection. *Goskompriroda* would be responsible for the environmental protection of Soviet natural resources. Beginning in 1989, a series of State annual reports on the state of the environment provided the first official account of environmental conditions and environmental protection efforts in the Soviet Union. The annual reports aimed "to promote the dissemination of verified environmental information, the mobilization of society's efforts to improve the environment, and rational use of natural resources, as well as the adopting of effective management decisions in this sphere."[28] The reports presented a compilation and synthesis of the work of a large number of environmentally related agencies and specialists and, indeed, became, as the 1999 report would state, a "unique" government document.

New legislation was promulgated under Mikhail Gorbachev and became the 1991 Law on Environmental Protection. The law specified: (1) a citizen's right to a healthy and safe environment; (2) a citizen's right to form environmental associations, to obtain information, and to seek legal redress for environmental change; (3) environmental responsibilities of the federal and other governmental levels; (4) environmental obligations of enterprises; (5) a state ecological examination system; (6) environmental liability; and (7) creation of an environmental funds system.[29] Other earlier laws, such as the 1982 Law on Air Protection, remained in force. Contradictions between new laws and existing laws would remain a hallmark of the reform period. Gorbachev also appointed a special presidential advisor to work on environmental issues. Aleksei Yablokov, a highly respected biologist and member of the Academy of Sciences, served as a highly visible presidential advisor into the beginning of the Yeltsin presidency.[30]

Press reports during this period carried more and more revealing details about environmental degradation that had taken place during the Soviet period. A Russian translation of *Ecocide in the USSR* by the Western specialists Murray Feshbach and Alfred Friendly reached a wide audience in Russia.[31] Many specialists in the Soviet Union criticized the book's conclusions as apocalyptic, but those same conclusions appeared to many others to be accurate. The activity of environmental NGOs grew significantly during this period. Such influential NGOs as the umbrella Socio-Ecological Union came into existence during this period. The Institute for Soviet-American Relations (ISAR, later renamed Initiative for Social Action and Renewal in Eurasia) opened an office in Moscow. Civil society grew from a small number of dissidents to a fledgling NGO community, as witnessed by the active presence of the Socio-Ecological Union throughout all of the republics of the Soviet Union by the end of 1992.

In 1992 a "polluter pays" principle was established, based on a system of norms of thresholds and relative multipliers. Ironically, Craig ZumBrunnen argued for the use of such economic levers as effluent charges back in 1975 at a joint Soviet-American conference on water quality under the auspices of the Nixon-Brezhnev Environmental Accord.[32] The call for such levers was made again in 1992 at the First Congress of the International Ukrainian Economic Association in Kiev, Ukraine.[33]

The new 1993 Constitution of the Russian Federation reinforced the importance and necessity of government environmental protection. Article 9 states that "the land and other natural resources are used and protected in the Russian Federation as the basis of the life and activity of the population inhabiting the corresponding territory."[34] Article 42 of the Constitution states that "everyone has the right to a healthy environment, accurate information about its conditions, and compensation for damage to health or property as a result of violation of environmental law."[35] Although budgetary funds were allocated to state environmental protection, inflation and recurring crises of nonpayment of transactions, consolidation of funds into budgets, and delays in fund transfers among jurisdictions meant that approved environmental projects were unlikely to reach fruition. Under increasingly difficult economic conditions, the Russian government would attempt to apply a market-based approach to its environmental protection efforts.

Goskompriroda and the Ministry of Finance together became responsible for implementing this pollution charge program. In 1994, an official new document, titled the "State Strategy of the Russian Federation on Environmental Protection and Sustainable Development," came to be the basis for many of the operative principles *of Goskompriroda*. A biennial action plan, the Government Action Plan for Environmental Protection for 1994 and 1995, also contained about 100 priority environmental measures.[36] At the same time, the Ministry of Natural Resources was created in 1996 on the previous foundation of the Committee of Geology and Natural Resource Use.[37]

Under the instituted Russian scheme, all polluting sources above a certain threshold became subject to a "base charge proportional to emissions or discharges of pollutants."[38] An accompanying system of Ecological Funds was established. The intention was to earmark the pollution charges collected for environmental protection only through the framework of the Ecological Funds. Pollution charges became the main source of revenue for those funds. The resources of the Ecological Funds were allocated on the principle of 10 percent to the federal level, and the remaining 90 percent to the regional and local level. Some conflicts emerged over access to those funds at the local level.[39] The total of revenue collected by Ecological Funds is estimated to have been about

US$ 2.2 billion for the entire period from 1992 to 1997.[40] Recently, Stig Kjeldsen has done a very thorough analysis of the strengths and weaknesses of the role of such charges in generating financial resources for the Federal Ecological Fund (FEF) and financing environmental protection in Russia.[41] His analysis reveals a number of problems with the current "pollution charge" scheme. Most notably these include: (1) problems with determining the magnitude of the charge; (2) charge levels being set too low; (3) exemptions based on environmental investments (the so-called Pollution Charge Exemption Scheme); (4) lack of incentives to reduce pollution due the practice of levying environmental charges for emissions with "maximum permissible levels (MPLs) and "Temporary compliance level (TCLs); (5) budget consolidation of "earmarked" ecological funds into the general budget of a given entity; and (6) the continuing growth of nonmonetary transactions in the overall Russian economy. On June 27, 2002, the State Duma gave the first reading of a chapter in the tax code. There had been some reports that enterprises were being refunded the "pollution charges" and that the Federal Ecologic Fund (FEF) was being abolished. This new tax legislation increases the overall number of taxes, and it specifically includes payments for the use of natural resources, including payment for the use of water objects, for the pollution of the environment, and for the use of forest resources.[42] Thus, it appears that the pollution charge scheme will continue. Much less certain is whether it will evolve into an effective environmental protection policy instrument or merely continue to function as a tax revenue-generating device!

The sharp industrial decline in the Russian economy in the mid-1990s meant that air pollution levels and drinking water quality were indeed improving. Some of the improvement came as the result of new air filter and water purification and treatment plants and some by the modest efforts to tackle the huge backlog of broken water mains and sewer pipes needing replacement. But the economic decline was far and away the largest determinant factor in terms of the decline in industrial pollution levels. Indeed, energy intensity levels (the amount of energy used per given level of economic output) increased in the 1990s. As Russia's economic numbers continued to decline, firms experienced sharper declines in production with only modest savings in their expenditure of energy. Nearly everywhere, financially strapped enterprises often opted to abandon compliance with environmental regulations as their first economizing measure.[43] As an in-depth environmental assessment of Moscow's environmental conditions has concluded, many of the expected improvements in environmental quality have not materialized.[44]

Goskomekologiia *and the Devolution of State Environmental Protection*

Government Decree Number 643 of May 26, 1997 replaced the Ministry of Environmental Protection with the State Committee on Environmental Protection (*Goskomekologiia*).[45] This loss in status of a state environmental protection agency came soon after Yeltsin's second election victory in the summer of 1996. The decision reflected a renewed interest in natural resource exploitation at the expense of a lowering of the stature of state environmental protection. *Goskomekologiia's* stated tasks were to: (1) implement and coordinate environmental policies; (2) develop environmental policy instruments; (3) implement state ecological examinations and inspections; (4) manage nature conservation; (5) establish and supervise environmental norms and standards; (6) prepare reports on the state of the environment and provide technical advice; and (7) manage the Federal Ecological Fund.[46] A final sphere of responsibility of *Goskomekologiia* involved international environmental cooperation.[47]

Goskomekologiia held offices at the republic, oblast, and krai levels. At the republic and oblast levels, *Goskomekologiia* maintained a relatively large amount of independence, often siding with local needs rather than federal-level preferences. In St. Petersburg, for example, the city-level administration for environmental protection was often at odds with *Goskomekologiia*, especially as concerned the distribution of resources of the regional Environmental Fund.[48] A number of other federal bodies also had jurisdiction over environmental protection issues. Those bodies were: (1) the Ministry of Public Health; (2) the Ministry of Emergency Situations; (3) the State Committee for Land Policy; (4) the State Committee for Fisheries; (5) the Federal Forestry Service; and (6) the Federal Service for Hydrometeorology.[49] The appointment of visible scientific bureaucrats, such as Victor Ivanovich Danilov-Danilian, the former minister of *Goskompriroda* (affectionately called Dan-Dan by some Russian environmentalists), as chairman of the new *Goskomekologiia,* provided some continuity from the former *Goskompriroda.* He continued to serve as chairman until the dissolution of *Goskomekologiia* in April 2000.

The record of *Goskomekologiia* was decidedly mixed. An evaluation of the success or insufficiencies of *Goskomekologiia*'s environmental protection record depends in large part on the local perspective. Reasonably well-trained and increasingly experienced ranks of thousands of inspectors had emerged by the end of the 1990s. Cases of bribery of those inspectors or other *Goskomekologiia* officials undoubtedly existed, but they appear to be the exception rather than the rule. Larger environmental projects appeared to be going ahead in the late 1990s, especially once the August 17, 1998, financial crisis had subsided. *Goskomekologiia's* offices communicated relatively openly and regularly with the environmental NGO community.[50] *Goskomekologiia* began to create World Wide Web–based environmental information resources.[51] Devolution of power within *Goskomekologiia* from the federal to the regional and local levels appeared to be providing both opportunities for creative environmental problem solving of environmental issues on the local level, but also for abuse and violations. In Bashkiria, for example, a dam was under construction in an area that was also considered part of a national park.[52] Elsewhere, *de facto* decentralization meant that decision-makers at the local level were "left to fill in the gaps" as they saw fit.[53] On balance, *Goskomekologiia's* work found both supporters and critics, but even its most vocal NGO critics would soon be appalled by the prospect of the agency's subsequent dismantlement.

Dissolution of Goskompriroda and Transfer to the Ministry of Natural Resources

Vladimir Putin's ascendancy to the presidency, first as acting president on December 31, 1999, and then by an overwhelming electoral victory three months later, resulted in a major retrogressive course-reversal for state environmental protection in Russia. Putin's self-proclaimed ideology of "strengthening of vertical power" sought to rein in the relative independence of the regions that had emerged in the 1990s.[54] Within two months after having assumed power as Russia's president by election, Putin issued Decree 867, which liquidated *Goskomekologiia* and transferred its responsibilities to the Ministry of Natural Resources. The May 17, 2000, decree also abolished the Federal Forestry Service and transferred its responsibilities to the same Ministry of Natural Resources. The 200-year-old Forestry Service had numbered about 100,000 employees.[55]

Putin's decision appeared to be a reaction to a number of events: (1) the devolution from centralized to decentralized management that had occurred within *Goskomekologiia*; (2) the August 1998 devaluation of the ruble, from which Russia's economy has begun only slowly to recover (though the devaluation is now widely viewed as a positive event from the point of view of economists); and, closely related, (3) renewed state support for an unencumbered exploitation of Russia's natural resources in order to revive Russia's economy as quickly as possible. The fallout from Decree 867 was almost immediate among Russia's nascent but increasingly cyber-networked environmental NGO community. Expressing disbelief, several NGO representatives clung to the point of view that the decision must have been made without

Putin's approval, and that the decision would soon be annulled. Such a large-scale elimination of a federal environmental protection agency appeared unprecedented for any industrialized country at the beginning of the twenty-first century. Despite some publicly expressed reservations by NGO representative about the objective record of achievements of *Goskomekologiia*, the former agency found an unlikely source of public support within the environmental NGO community.

The Socio-Ecological Union, Russia's largest umbrella organization of environmental NGOs, decided to collect the requisite number of citizen signatures (2 million by Russian law) in order that an officially sanctioned national referendum be conducted on three environmental questions, two of which were directly related to Decree 867. The three questions proposed for the referendum, and for which a signature drive was launched immediately, were: (1) "Do you agree with the decision to abolish Russia's state environmental protection agency (*Goskomekologiia*)?"; (2) "Do you support the import of nuclear wastes from abroad into Russia?" (a common practice during the Soviet period; this practice had been stopped in the early 1990s by law); and (3) "Do you support the abolition of the federal forest agency?" Question 2, in particular, while not directly related to Decree 867, was strategically included as one of the three questions on the signature drive for the proposed referendum. The Russian environmental NGO community anticipated that such a question would elicit an unambiguously negative reaction among the Russian public.[56]

From May through September 2000, representatives of more than 100 environmental NGOs in more than 50 cities in Russia worked to publicize the signature drive. They organized petition stands at city center locations and at specially organized events, and in general worked tirelessly to collect the required number of signatures for conducting an official referendum at the national level. The effort proved to be a well-coordinated and sustained one, and by the end of September 2000, almost 600,000 more signatures had been collected than the requisite 2 million signatures for a national referendum to be approved and conducted. But upon a technical review of the signatures by the Central Election Committee in Moscow (that review was conducted in Moscow as well as locally), the Central Election Committee made a concerted and swift effort to eliminate signatures for technical reasons.[57]

On November 29, 2000, the Central Election Committee ruled on the basis of incorrectly abbreviated addresses and a number of seemingly innocuous technical points that an insufficient number of signatures had been collected (i.e., less than 2 million) for a national referendum to be held on reinstating a state environmental protection agency. An official court appeal by the Socio-

Ecological Union resulted in an officially stated reaffirmation of the Central Election Committee's original finding that 600,000 votes were missing.[58]

In a further blow to the organizers of the original signature drive, in June 2001 President Putin signed a decree to permit the import of nuclear waste into Russia for the reported purpose of reprocessing. A reported $20 million would be earned from this reprocessing, though the details of the exact source of that revenue have never been publicly released. The Ministry of Atomic Energy argued that such funds were required so that the Ministry could clean up existing nuclear waste sites in Russia, a conclusion that has been viewed as largely spurious among environmental specialists in Russia and the West.[59]

The Russian government ostensibly sought to find some common ground with the Russian environmental NGO community in the fall of 2001 when it conducted a highly publicized "Civic Forum" with NGO representatives invited from throughout Russia. Putin addressed the representatives in person, voicing support for their work. But it would appear that his pledge of support was only partially genuine or at least fleeting, as no follow-up activities have been conducted since the Forum. As has been noted, "the state is in no shape to support public movements, and moreover it has little interest in encouraging them."[60]

The "Strengthening of Vertical Power" within the Ministry of Natural Resources

This most recent period has widely been viewed as one of the "de-greening" (*de-ekologizatsiia*) of the Russian state.[61] Any hope that the newly recreated Ministry of Natural Resources might retain any substantial state environmental protection appears to be largely without justification. *Goskomekologiia*'s previous ranks of inspectors, reduced significantly in size from their original numbers, have become a subordinate part of the Ministry of Natural Resources. The loss of expertise from the former *Goskomekologiia* will likely be long-lasting. "We have witnessed a sudden and nearly complete collapse [of state environmental protection], marked by a mass exodus of staff, problems with document circulation, and silence in response to official inquiries," one NGO representative has concluded.[62] Further suggestions have been made that the new Ministry of Natural Resources has been designed to orchestrate the upcoming privatization of forestlands to benefit the appropriate oligarchs.

In some exceptional cases, city administrations have been successful in retaining their city-level administration for environmental protection. For example, in the case of St. Petersburg, the administration-level environmental agency has been recently renamed the Administration for Environmental

Safety and Natural Resource Use. Despite its new name, this agency appears to have retained its environmental protection responsibilities in full. But such positive examples appear to be the exception rather than the rule. Further attempts to create the outward appearance of retaining the trappings of a state environmental protection agency seem to have been lost on the Putin government. State-sponsored environmentally sensitive/threatening initiatives, such as the recently completed Baltic Pipeline System or oil extraction development on Sakhalin Island and offshore in its coastal fishing grounds, have instead not surprisingly met with no resistance or significant interference from within the Ministry of Natural Resources from the point of view of environmental protection. As one Russian commentator has observed: "There simply is no environmental policy in Russia—the existing policy could actually be construed as intending to destroy environmental policy."[63]

If any positive developments have occurred in terms of state environmental protection since April 2000, it may be in terms of the improvement of accessibility to some basic environmental information resources within the Ministry of Natural Resources. The annual reports on the "Status of the Environment" are readily available online at the Ministry's website. One of the major drawbacks of the annual reports produced by the oblast-level offices of *Goskomekologiia* had been their very small press runs. Also, official environmental publications are with increasing frequency being made available on oblast-level websites. The ministry also funds two newspapers with environmental coverage, *Prirodno-resursnye Vedomosti* and *Ekologicheskaia Gazeta Spasenie*, though each has an admittedly government rather than activist perspective. Those public officials who did not regret the passing of the former *Goskomekologiia* remain optimistic that a better-financed agency, the Ministry of Natural Resources, will provide more opportunities for investment in environmental infrastructure such as wastewater and purification plants than *Goskomekologiia* had in the past. But such optimists are also admittedly few.

Human Rights and Environmental Whistle-Blowers in Russia

Human rights issues remain critical for Russian environmental activists, as the cases of Grigorii Pasko (a naval journalist accused of revealing naval secrets concerning dumping in the Sea of Japan) and Igor Sutiagin (accused of spying and transferring state secrets to Western government representatives, though he has demonstrated that his only sources were from the public record) continue to demonstrate. The acquittal of Alexandr Nikitin, after more than five years of court proceedings and delays and a one-year jail term, appears to be an exception that was made for a Russian whistle-blower under the lobbying pressure and publicity campaign successfully aimed at the court of world opinion. As has been noted, Russian courts do not have a good record of independence.[64] Incidents of employee firings at nuclear power plants and other environmentally sensitive sites continue to occur regularly as whistle-blowers attempt to bring environmental risks to the light of the public. Russian environmental NGOs' almost inevitable reliance on foreign financial assistance (especially under conditions of active opposition to so many of the Russian government's current policies) continues to come under attack from the highest levels of the Russian government. From the point of view of Western governments, however, this support is one of the best possible peace dividend investments.

Alexandr Nikitin, a former naval officer based in Murmansk, drew the wrath of the Russian military establishment for his co-writing of a report for the Norwegian NGO Bellona on the topic of nuclear hazards from the Soviet and Russian navy in the Barents Sea region. Nikitin was arrested in February 1996 and held in solitary confinement for fourteen weeks. After more than a year in prison for alleged spying and release of state secrets to a foreign government, he was released and drew international attention to human rights abuses on Russian whistle-blowers. Nikitin was later fully absolved of his accusations, but only after two years of highly public trials that to many viewers revealed to what extent some authorities would go in an effort to conceal environmental information if it was considered even remotely related to militarily sensitive information and activities. The Russian Supreme Court eventually heard his case. Niktin's conviction created an outrage both internationally and in Russian environmental NGO circles. Nikitin's lawyers, engaged by Bellona, were seen as having played a critical role in Nikitin's acquittal. The fate of another Russian whistle-blower, Grigory Pasko, has been less fortunate. Pasko worked as an investigative journalist for the newspaper of the Russian Pacific Fleet, "Boyevaya Vakhta," where he focused on nuclear safety issues. He was arrested by the Russian Security Service (FSB) in November 1997 and accused of committing treason through espionage when working with Japanese journalists. The Court of the Pacific Fleet acquitted Pasko of the treason charges in July 1999 and released him under a general amnesty. Yet the Military Collegium of the Russian Supreme Court reversed the verdict in November 2000 and sent the case back to the Pacific Fleet Court for a retrial. Pasko was next sentenced to four years of prison in December 2001. Whether or not the Russian Supreme Court will hear his appeal remains unresolved, and a decision was expected in June 2002. Both cases came to be highly publicized in Russia and have been viewed as critical indicators of the tolerance of the Russian government toward environmental whistle-blowers in general.[65]

Conclusions

The cycle of the strengthening and weakening of Russia's state environmental protection during the 1990s has had some lasting effects. It can be strongly argued that in the face of opposition, the Russian environmental community has emerged stronger. Accordingly, under conditions of a clearly identifiable opponent, the environmental community's efforts might be more effectively targeted and deployed today. The experience of the summer 2000 signature drive has also likely forever changed the Russian environmental community. They proved to themselves that they could coordinate citizenry political action on a national scale. Even if their first attempt did not meet with success, they did make a serious statement of their views to the government. Russian environmental NGOs recently have also been able to begin to recruit some of the environmental specialists who previously worked with the former *Goskomekologiia*. Such levels of expertise assist greatly in counteracting the prevalent image of NGO representatives as uninformed and poorly trained.

It is possible that the recent dismantlement of state environmental protection may be limited in the realm of international environmental cooperation, due to the fact that Russia is the co-signer of a large number of international initiatives and bi- and multi-lateral environmental programs.[66] Indeed, Alexei Yablokov, who is now head of the Center for Environmental Politics in Moscow, in March 2001 during a visit to Washington called for a more active stance on the part on the U.S. government to assist in blocking the import of nuclear waste into Russia in the future, as the U.S. controls the vast majority of those wastes on a worldwide scale. Alexander Nikitin's recent efforts in the human rights realm have included the active solicitation of support from foreign NGOs to draw attention to human rights abuses in Russia. His work is an indication not only of the continuing opposition the environmental NGO community in Russia confronts in the face of the Russian government, but also of the Russian NGO community's continuing dependence on the outside world for both financial and moral support.

The environmental policy challenge that the new Russia faces in terms of environmental degradation and continuing environmental disruption remains considerable and serious. Russia must not only address new environmental challenges, but also work to repair the widespread ecological damage inflicted by a particularly destructive recent past.[67] While a sustained economic upturn in Russia might lead to legislative and bureaucratic reform in state environmental policy in the future, the current status of state environmental protection in Russia appears to be only marginally improved over that of the end of the Soviet period. Despite significant results during the 1990s in terms of state intervention to improve Russia's natural environment, the period of seemingly genuine concern for the environment would appear to have passed. Indeed, one well-known environmental NGO representative has called the Ministry of Natural Resources' steps in 2002 to be "reminiscent of the 1930s-1950s" in terms of its (Stalinist-like) style of leadership.[68]

Current developments would appear to prevent the likelihood of the re-emergence and strengthening of state environmental protection in Russia any time soon. The Ministry of Natural Resources has a mandate to decide any environmentally controversial question on the side of increased natural resource extraction and profitability. Indeed, it would be difficult to envision a government-sponsored project that might be stopped by the Ministry of Natural Resources on environmental grounds, given the stated and express purpose of the Ministry to prioritize the extraction and use of natural resources for Russia's at least short-term economic advantage. Its purpose to prioritize economic development is unambiguous. At the same time, as long as economic policies and incentives at the state level give the appearance of promoting the practical challenges of Russian citizens to live a "normal" life, a strengthening of state environmental protection policies and practices will likely be seen as a luxury for Russia's leaders for a long time to come. Only as more and more Russian citizens fervently come to appreciate their well-being not only in terms of their material wealth, but also in terms of the health of their children and of the recreational opportunities of an unpolluted environment, will the Russian state in future feel obliged to adopt positive environmental protection policies.

Notes

1. See S. A. Ushakova and Ia. G. Katsa, eds., *Ekologicheskoe sostoianie territorii Rossii* (Moscow: Academia, 2001), pp. 20–33.

2. For example, ibid., pp. 62–104.

3. Yegor Belorus, "Moscow's Drawbacks: The Capital Is Like Another Chernobyl," available at www.pravda.ru, July 2, 2002, translated by Dmitry Sudakov.

4. For example, Ushakova and Katsa, eds., *Ekologicheskoe sostoianie territorii Rossii*, pp. 104–109. Also, see map *18 Radiatsionnie zagriaznenie* located between p. 96 and p. 97.

5. For example, see M. Feshbach and Alfred Friendly, Jr., *Ecocide in the USSR: Health and Nature under Siege* (New York: Basic Books, 1992).

6. Andrew R. Bond, "Air Pollution in Noril'sk: A Soviet Worst Case?" *Soviet Geography*, vol. 25, no. 9 (November 1984), pp. 665–680.

7. A. V. Yablokov and V. I. Danilov-Danil'yan, eds., *Gosudarstvennyi doklad: O sostoianii okruzhaiushcheyi sredy Rossiiskoi Federatsii v 1991 godu* (Moscow: Goskompriroda, 1992), p. 4, appendix.

8. Ushakova and Katsa, *Ekologicheskoe sostoianie territorii Rossii*, p. 9.

9. Ibid., p. 13.

10. Ibid., p. 17.

11. Craig ZumBrunnen, "A Review of Soviet Water Quality Management: Theory and Practice," Chapter 13 in *Geographical Studies on the Soviet Union: Essays in Honor of Chauncy D. Harris*, Research Paper No. 211, edited by Roland Fuchs and George Demko (Chicago: The University of Chicago, 1984), pp. 257–294; and Boris Komarov, *The Geography of Survival: Ecology in the Post-Soviet Era* (Armonk, N.Y: M.E. Sharpe, 1994).

12. Denis J. B. Shaw, *Russia in the Modern World: A New Geography* (Oxford: Blackwell, 1999), p. 133.

13. Craig ZumBrunnen, "Institutional Reasons for Soviet Water Pollution Problems," *Proceedings of the Associations of American Geographers*, vol. 6 (1974), pp. 105–108; and ZumBrunnen, "A Review of Soviet Water Quality Management: Theory and Practice," pp. 270–273.

14. See Judith Thornton and Andrea Hagan, "Russian Industry and Air Pollution, What Do the Official Data Show?" *Comparative Economic Studies*, vol. 34, no. 2 (Summer 1992), p. 19.

15. See Jonathan Oldfield, "Environmental Impact of Transition—A Case Study of Moscow City," *Geographical Journal*, vol. 165, no. 2 (1999), pp. 222–231.

16. Komarov, *The Geography of Survival*, p. 56.

17. James H. Bater, *Russia and the Post-Soviet Scene: A Geographical Perspective* (New York: John Wiley and Sons, 1996), p. 316.

18. Craig ZumBrunnen, "The Lake Baikal Controversy: A Serious Water Pollution Threat or a Turning Point in Soviet Environmental Consciousness?" in *Environmental Deterioration in the Soviet Union and Eastern Europe*, edited by Ivan Volgyes (New York: Praeger, 1974), pp. 90–114.

19. Loren R. Graham, *The Ghost of the Executed Engineer: Technology and the Fall of the Soviet Union* (Cambridge, MA: Harvard University Press, 1993).

20. For example, see Peter Rutland, "Sovietology: Who Got It Right and Who Got It Wrong? And Why?" in *Rethinking the Soviet Collapse: Sovietology, the Death of Communism and the New Russia*, edited by M. Cox (London and New York: Pinter, 1998), pp. 32–50; and Oldfield, "Environmental Impact of Transition—A Case Study of Moscow City," pp. 222–231.

21. N. A. Gladkov, "Bogatstva prirody: zabotlivo okhraniat', razumno ispol'zovat'. Vosstanavlivat' i umnnnnozhat'," *Priroda*, no. 2 (1962), p. 6.

22. Igor Petryanov, "Quo Vadis," *Soviet Life* (November 1970), p. 53.

23. G. Khromushin, "Is Technology to Blame?—Who Benefits from Concepts of an Irreversible Ecological Crisis?" *Current Digest of the Soviet Press*, vol. 31, no. 37 (October 10, 1979), p. 15.

24. ZumBrunnen, "Institutional Reasons for Soviet Water Pollution Problems," pp. 105–108.

25. For example, see Andrew R. Bond and Matthew J. Sagers, "Some Observations on the Russian Federation Environmental Protection Law," *Post-Soviet Geography*, vol. 33, no. 7 (September 1992), pp. 463–474; Komarov, *The Geography of Survival*, pp. 3–13; D. J. Peterson and Eric K. Bielke, "The Reorganization of Russia's Environmental Bureaucracy: Implications and Prospects," *Post-Soviet Geography and Economics*, vol. 42, no. 1 (January–February 2001), pp. 65–76; Craig ZumBrunnen and Nathaniel Trumbull, "Obstacles and Opportunities to the Establishment of an Environmental Information Network in Northwest Russia," *Journal of Urban and Regional Development Research*, vol. 8, no. 1 (2000), pp. 38–58; and Craig ZumBrunnen and Nathaniel Trumbull, "An Emerging Northwest Russia Environmental

Information Network: IT Capacity Building for Environmental Protection and Sustainable Development," *NETCOM*, vol. 15, nos. 3–4 (2001).

26. T. Saiko, "Environmental Crises: Geographical Case Studies in Post-Socialist Eurasia" (Harlow, England: Prentice Hall, 2001), p. 61.

27. A. Knorre, "The Rise and Fall of Environmental Protection as a National Security Issue," *Russia's Fate through Russian Eyes: Voices of the New Generation* (Boulder: Westview, 2001), p. 291.

28. For example, see Goskomekologiia, *O sostoianii okruzhaiushchei pridronoi sredy Rossiiskoi Federatsii v 1999 g.* (Moscow: Goskomekologiia, 2000).

29. OECD, *Environmental Performance Reviews. Russian Federation* (Paris: Organization for Economic Cooperation and Development, 1999), p. 45; and Bond and Sagers, "Some Observations on the Russian Federation Environmental Protection Law," pp. 463–473.

30. Peterson and Bielke, "The Reorganization of Russia's Environmental Bureaucracy: Implications and Prospects," pp. 65–70.

31. Feshbach and Friendly, Jr., *Ecocide in the USSR*.

32. Craig ZumBrunnen, "Vliianie geografo-ekonomicheskikh faktorov na sistemy upravleniia kachestvom vody," *Ispol'zovanie matematicheskikh modeley dlia optimizatsii upravleniia kachestvom vody: Trudy Sovetsko-Amerikanskogo simpoziuma, tom I* (Leningrad: Gidprometeoizdat, 1979), pp. 186–216.

33. Craig ZumBrunnen, "Mechanisms for Environmental Quality Management: Framework for Application in Ukraine," in *Ekonomika ukraini: minule, sucasne i maybutnh—The Economy of Ukraine: Past, Present and Future*, Proceedings of First Congress of the International Ukrainian Economic Association, edited by George Chuchman and Mykola Herasymchuk (Kiev: Ukrainian Academy of Sciences, Institute of Economics, 1993), pp. 168–183.

34. OECD, *Environmental Performance Reviews. Russian Federation*, p. 45.

35. Knorre, "The Rise and Fall of Environmental Protection as a National Security Issue," p. 294.

36. OECD, *Environmental Performance Reviews. Russian Federation*, p. 51.

37. N. D. Sorokin, ed., *Okhrana okruzhaiushchei sredy, pripodopol'zovanie i obespechenie ekologicheskoi bezopasnosti v Sankt-Peterburge za 1980–2000 gody* (St. Petersburg: Administratsiia Sankt-Peterburga upravlenie po okhrane okruzhaiushchei sredy, 2000), p. 30.

38. OECD, *Environmental Performance Reviews. Russian Federation*, p. 142.

39. OECD, *Environmental Financing in the Russian Federation* (Paris and Washington, DC: Organization for Economic Cooperation and Development, 1998).

40. OECD, *Environmental Performance Reviews. Russian Federation*, p. 147.

41. Stig Kjeldsen, "Financing of Environmental Protection in Russia: The Role of Charges," *Post-Soviet Geography and Economics*, vol. 41, no. 1 (January–February 2000), pp. 48–62.

42. Ria Oreanda, "Taxes, Economic Press Review: Taxes," *Delovoy Peterburg*, June 28, 2002.

43. Andrew R. Bond, "Environmental Disruption During Economic Downturn: White Book Report," *Post-Soviet Geography*, vol. 34, no. 1 (January 1993), p. 75.

44. Oldfield, "Environmental Impact of Transition—A Case Study of Moscow City," pp. 222–231; and Victoria R. Bityukova and Robert Argenbright, "Environmental Pollution in Moscow: A Micro-Level Analysis," *Eurasian Geography and Economics*, vol. 43, no. 3 (April–May 2002), pp. 197–215.

45. Sorokin, ed., *Okhrana okruzhaiushchei sredy, pripodopol' zovanie i obespechenie ekologicheskoi bezopasnosti v Sankt-Peterburge za 1980–2000 gody*, p. 32.

46. OECD, *Environmental Performance Reviews. Russian Federation*, p. 44.

47. Peterson and Bielke, "The Reorganization of Russia's Environmental Bureaucracy: Implications and Prospects," p. 74.

48. Nat Trumbull interview with Anatoly Baev in St. Petersburg, Russia, December 1999.

49. OECD, *Environmental Performance Reviews. Russian Federation*, p. 52.

50. Craig ZumBrunnen and Nathaniel Trumbull, "Obstacles and Opportunities to the Establishment of an Environmental Information Network in Northwest Russia," pp. 38–58.

51. Information obtained from discussions held by authors at UNEP/GRID-Arendal and Swedish EPA-sponsored workshop entitled "Strengthening Information Management and Reporting on the Environment and Sustainable Development for North-West Russia and Belarus," held in Moscow, April 12–14, 2000. Also see http://www.grida.no/enrin/nwrussia/index.htm.

52. Knorre, "The Rise and Fall of Environmental Protection as a National Security Issue," p. 289.

53. H. I. Glushenkova, "Environmental Administrative Change in Russia in the 1990s," *Environmental Politics*, vol. 8, no. 28 (Summer 1999), p. 161.

54. V. Putin, V., "Vystuplenie Prezidenta Rossiiskoi Federatsii V.V. Putina na zasedanii Gosudarstvennogo soveta Rossiiskoi Federatsii," November 22, 2000. Available at: http://president.kremlin.ru:8104/events/105.html.

55. Peterson and Bielke, "The Reorganization of Russia's Environmental Bureaucracy: Implications and Prospects," pp. 65–76; and Craig ZumBrunnen and Nathaniel Trumbull, "Obstacles and Opportunities to the Establishment of an Environmental Information Network in Northwest Russia," pp. 38–58.

56. Information gleaned from interviews conducted by Nathaniel Trumbull in St. Petersburg, Russia, and other cities of northwest Russia during the summer of 2000.

57. Nathaniel Trumbull and Craig ZumBrunnen, "The Debilitating Transformation of Environmental Protection Institutions in Russia," paper presented at AAG Annual Meetings, Los Angeles, CA, March 20, 2002.

58. E. Shvarts, "What Is Happening in the Ministry of Natural Resources," *Russian Conservation News* (Winter 2002), pp. 4–5.

59. Ibid.

60. Knorre, "The Rise and Fall of Environmental Protection as a National Security Issue," p. 293.

61. Craig ZumBrunnen and Nathaniel Trumbull, "Obstacles and Opportunities to the Establishment of an Environmental Information Network in Northwest Russia," p. 64.

62. Shvarts, "What Is Happening in the Ministry of Natural Resources," pp. 4–5.

63. Knorre, "The Rise and Fall of Environmental Protection as a National Security Issue," p. 286.

64. Denis J. B. Shaw, *Russia in the Modern World: A New Geography*, p. 142.

65. Nathaniel Trumbull and Craig ZumBrunnen, "The Debilitating Transformation of Environmental Protection Institutions in Russia," paper presented at AAG Annual Meetings, Los Angeles, CA, March 20, 2002; and discussions with Aleksandr Nikitin during his visit to Seattle in February 2002.

66. Peterson and Bielke, "The Reorganization of Russia's Environmental Bureau-cracy: Implications and Prospects," p. 74.

67. Ilmo Massa and Veli-Pekka Tynkkynen, eds., *The Struggle for Russian Environmental Policy* (Helsinki: Kikimora Publications, 2001).

68. Shvarts, "What Is Happening in the Ministry of Natural Resources," pp. 4–5.

For Further Reading

Philip R. Pryde, *Environmental Resources and Constraints in the Former Soviet Republics* (Boulder: Westview Press, 1995).

OECD, *Environmental Performance Reviews. Russian Federation* (Paris: Organization for Economic Cooperation and Development, 1999).

W. U. Chandler, *Energy and Environment in the Transition Economies: Between Cold War and Global Warming* (Boulder: Westview Press, 2000).

Ilmo Massa and Veli-Pekka Tynkkynen, eds., *The Struggle for Russian Environmental Policy* (Helsinki: Kikimora Publications, 2001).

D. J. Peterson and Eric K. Bielke, "The Reorganization of Russia's Environmental Bureaucracy: Implications and Prospects," *Post-Soviet Geography and Economics*, vol. 42, no. 1 (January–February 2001), pp. 65–76.

T. Saiko, *Environmental Crises: Geographical Case Studies in Post-Socialist Eurasia* (Harlow, England: Prentice Hall, 2001).

Russian Environmental Digest (newsletter in English). Subscription information for the Russian Environmental Digest: To subscribe, write to <majordomo@teia.org> with "subscribe redfiles" in message body.

Index

Asian-Russian relations. *See* Russian-Asian relations
Aspin, Les, 85
Association of South East Asian Nations, 70
Astrakhan, 51, 58
Atomic Energy Agency. *See* International Atomic Energy Agency
Aushev, Ruslan, 129
Azerbaijan, xxiv, 48, 49, 67, 68, 69, 159, 211
 conflict in, 40, 44, 60

Baker, III, James A., 81, 96
Baltic states, 11, 49, 210–11
Bashkiria, 128
Bashkortostan, 51, 127, 133, 137
Belaia Vezha accords, 47
Belarus, 4, 8, 43, 46, 47, 48, 80, 210
Belarus-Russian Union, 60
Bennett's law, 239
Berezovsky, Boris, 170–71
bin Laden, Osama, 51–52, 65, 66, 72
Black Sea Forum, 60
Blair, Tony, 95, 96
Blue Stream pipeline, 69–70, 74
Bolshevik Revolution, 142, 143–44, 201
Border disputes, Russian-Chinese, 3, 27–28
Border Guards, 4, 8, 15
Bosnia, 10, 11
Brzezinski, Zbigniew, 96
Bulgaria, 11
Bush, George H.W., 81
Bush, George W., 52, 53, 79, 89–90, 92, 93, 96–97

Caspian basin. *See* Caspian region
Caspian Basin Initiative, 69
Caspian Council, 69
Caspian Finance Center, 69

Caspian Five, 60
Caspian Littoral Agreement, 68
Caspian region, 60, 63, 66–70
Caucasus, xxv, 58–59, 62, 85, 109, 111, 117. *See also* Armenia; Azerbaijan; Georgia; North Caucasus; Russian-Eurasian relations
Caucasus Four, 60
Center-periphery relations, xxiii, 49, 123–40
 budgetary federalism, 133–35
 federal districts, 124–26, 138
 Federation Council reform, 135–36, 138, 173
 governors and republic presidents, 125–127, 128–33, 134, 136–38
 laws, resolving differences between regional and federal, 127–28
 presidential representatives, 124–26, 127
 under Putin, 49, 124–38
 under Yeltsin, 123–24, 127, 137
Central Asia, xxiv, xxv, xxvii, 32–33, 46, 51, 52, 58–59, 60, 85, 210
 Russian-American relations in, xxv, 46, 67, 68, 69, 72–73, 74
 See also Kazakhstan; Kyrgyzstan; Russian-Eurasian relations; Tajikistan; Turkmenistan; Uzbekistan
Central Asian Cooperation Organization, 60
Central Asian Union, 60
Charter of Russian-American Partnership and Friendship, 81
Chechnya, xxiv, 44, 54, 58, 67, 69, 111, 123
 conflict in, 5, 8, 9, 16, 40, 48–50, 51–52, 60, 124, 170–71
 and terrorism, 9, 15, 51–52, 72, 91, 124, 170–71